"*I strongly feel that for anyone looking forward to join Software industry as a development engineer or a test engineer, this book would prove to be a valuable resource. What I liked most is that not only the authors have done a nice job of bridging the gap between campus and corporate but have done so while keeping the content concise, to the point and interesting.*"

~ Ajay Nema, Founder, WhistleTalk

"*This book is perfect starting point for people who are entering the corporate world. Gives a a complete view of software life cycle. Walk you through good code with detailed code samples, explaining interesting debugging tips and tricks and testing methodologies. If you are a experience software engineer, you can make use of this book to put some of your own practice into words, to explain it to other people in your team. You will probably learn a few new things as well. In all a good book that will definitely help you grow your career.*"

~ Akshat Vig, Software Development Engineer, Amazon

"*Ashish is a true genius when it comes to coding, algorithms, ideation or cracking the TOUGH NUTS open. But, after reading this work of his, you will be convinced that he is blessed when it comes to writing and presenting his genius to others as well. A SUPER STAR SOLUTION for every fresher wanting to make his way into & UP in the industry.*"

~Sujit Lalwani, Author of 'Life Simplified!'

T0121033

"It was my pleasure to get an opportunity and go through the book which takes through the complete life-cycle of the software development phases. All engineers are accustom to the software development phases and almost all graduate courses covers various flavors of the software engineering. However the most challenging part of the software engineering is to relate with day-today usage in the software industry.

Even when I graduated, I was completely unaware of the essence of the software engineering till I experienced and gained industry experience. I didn't have much clue about some of the software industry's best practices such as logging, debugging tools, version control, defect tracking etc. It is extremely joyful to read such a book, which fills the gap between academic curriculum and industry experience.

The book not only talks about the industry's best practices but also provides many references to various tools and technologies which will take the readers to few notches up from best coder to the smartest developer. My sincere congratulations and best wishes to the authors for their outstanding work."

~ Satyendra Tiwari, Senior Architect, Citrix

HELLO WORLD

Student To Software Professional - A transformation Guide

Authored by:

Ashish Vaidya,
Pankaj Pal

PARTRIDGE
A Penguin Random House Company

To order additional copies of this book, contact
Partridge India
000 800 10062 62
www.partridgepublishing.com/india
orders.india@partridgepublishing.com

Preface

Cheers! You are going to be a *professional*.

The real world, as some people call it, awaits you. Soon you'll be working with people of distinctive caliber, contributing to projects worth a fortune, and handling stuff in a responsible position where you can make a difference. Sounds great! Isn't it?

With similar anticipations, brimming with excitement and enthusiasm, and dressed in well-ironed brand new formals, someone just like you walked in through the main entrance of an MNC on his first day. Ready to impress everyone with his coding prowess, our friend somehow managed to sit through the five-day induction program, attributing to the presence of this one beautiful girl in the room. After getting introduced to teammates, project, cubicle and equipment, our friend was assigned his first task- to understand the given project and make a small enhancement. The code was mostly written in familiar C and some portion in not-so-familiar Perl. "Wouldn't be too difficult", that's what he exclaimed to his new good-looking friend.

However, the enormous code spanning across hundreds of files, directories and sub-directories started overwhelming him. Perl was something new to be learnt but even the C code looked quite different from what was practiced in college. Forget about catching up with the girl, our friend was already staying late, feeling feeble as he was trying to make sense out of the vast code and learn Perl basics from a book.

The intentions are not to intimidate you but demonstrate with help of a small real life example how the things are different in a corporate environment. Most of the young coders entering the corporate world find themselves in such not-so-pleasant situations and what gets affected most is their confidence. A jolt to confidence in the initial days may degrade a potential stud into an average dud who just keeps complaining of being unhappy with the job.

There is a huge gap between how things are taught and learnt in college and the way they are applied in industry. For example, one hardly considers aspects like *scalability, readability, maintainability, portability, reusability, security concerns, performance, space and time complexity* or even *readability* while writing code in college. For most of the students, the main objective is just to display the correct output. After all, that's what you would be evaluated upon. Why these italicized words and more such terms did not seem important then is also due to the fact that the stakeholders are few- mostly only you, and scope of the usage of your code is also very limited- mostly only to yourself.

However, the scenario is different in industry. Characteristics mentioned above become very important when you take the responsibility of writing code working at a reputed organization whose image is at stake with every single line of code being added to its repository. Also, the development here is a continuous ongoing process having many stakeholders and many programmers contributing to it. Therefore, following coding guidelines and certain processes to do things in a streamlined manner becomes necessary.

You must have now understood that working in a professional environment is a different ball-game altogether. This book aims to help you prepare for the same, at least from a technical perspective.

This book doesn't claim to teach something revolutionary. In fact, everything you learn by reading this book can be learnt over years by experience. The latter method for acquiring this knowledge can however be tedious, painful, time-consuming and sometimes discouraging. Instead of toiling hard to reinvent the wheel, you may choose to grasp the readily available learning served here, and be ready for the industry right from day one.

This book is not entirely language specific but it uses C programming environment for explaining the concepts and examples. C was chosen because it is taught and learnt as a part of curriculum in almost all engineering colleges. Therefore, even the students who come from a different engineering stream but choose to build their career in the field of software will find it familiar and easy to understand. This, by no means, restricts one to extend and apply the knowledge gained to other programming environments.

This book is divided into 10 straightforward chapters written in a sequential manner where each one focuses on a different aspect. You'll get to learn how you can add professional touch to the aspects you already know, e.g. *writing code* and about the stuff you may not already be familiar with e.g. *Product Shipment Cycle*. Each chapter concludes with a Chapter Review that summarizes key points for your convenience. Keep an eye on *Tool-tips* and *Notes* in the chapter. You'll definitely find them useful. A list of appendix is provided to augment your understanding of certain topics which require a bit of extra learning.

To keep the topics crisp, concise and easy to understand, we have limited the details to a decent level which would be more than sufficient to prepare you. Later, you may want to go deep and explore more on certain topics from other resources and we would encourage you to do that. Some good books for you to explore as you advance in your career are mentioned on the last page. After all, to keep growing, you need to keep learning.
Wish you a great career ahead.

Best Wishes,
Ashish Vaidya
Pankaj Pal

Acknowledgements

Writing a book is a massive collaborative effort and no book is complete without the mention of several helping hands who contribute to it directly or indirectly. That said an author's work gets as good as the support he receives from his family, friends and well-wishers.

First and foremost, we would like to thank the supreme creator who gave us power to realize our passion for writing, believe in ourselves, pursue this project and complete it.

Throughout this project, various people gave inputs and helped improve our work. We thank them all and would like to specifically mention of *Ajay Nema, Co-Founder WhistleTalk; Satyendra Tiwari, Senior Architect, Citrix; Akshat Vig, Software Development Engineer, Amazon; Sujit Lalwani, Founder and Director, Inspiration Unlimited* who also took out their valuable time to review our work. We also acknowledge *Joseph Lobo* for the nice cover design.

Pankaj Pal would like to thank his dear wife, Komal for taking care of all the daily chores so that he could focus entirely on the writing. He thanks his colleagues, friends & cab-mates for the insightful discussions without which the book could never have been completed.

Ashish would like to thank his mother Raj Vaidya, father Lokesh Vaidya and brother Akhil Vaidya for always being there. His friends Vivek Agrawal, Sankar P, and all the members of the team at Inspiration Unlimited deserve a mention for being continuous sources of motivation. He also appreciates the support of all his friends, and well-wishers.

Last but not the least, we gratefully acknowledge the Partridge Publishing for publishing this book.

Contents

Writing code is elemental for a Software Engineer; Writing Good code is indispensable trait for a Good Software Engineer. This chapter covers what differentiates good code from code and makes life easier for the one who codes, one who reviews and one who maintains the code. The topics focus on Useful Programming Practices and Guidelines, many tips with examples and common mistakes made by programmers. Also, the importance and usefulness of professional code editors and their usage is also covered.

Writing code is elemental for a Software Engineer; Writing Good code is indispensable trait for a Good Software Engineer. This chapter covers what differentiates good code from code and makes life easier for the one who codes, one who reviews and one who maintains the code. The topics focus on Useful Programming Practices and Guidelines, many tips with examples and common mistakes made by programmers. Also, the importance and usefulness of professional code editors and their usage is also covered.

What good is code if it can't be compiled or its output can't be checked? The title might seem elusively simple but the inherent process is something that has evolved over years to simplify things for large code bases where files are spanned across multiple directories and subdirectories. Also, output viewing is something that is different for daemons when compared to applications that just execute and exit. This chapter covers concepts like compilation, linking, Makefiles, libraries, appropriately using the logs to view the output of processes and daemons.

Every newbie and even some experienced people struggle to go through and understand someone else's code. Moreover frustration level to do so increases proportionally with the size and span of codebase. This is something that one usually learns only through experience over a period of time. This chapter aims to help one explore and understand a completely unknown code base without going through it line by line.

Debugging is arguably the most important skill that a software engineer needs. Using a debugger and trace the issue by walking through the program line by line is the something that generally comes to mind when one hears of this term. However, there is lot more to debugging. This chapter covers the debugging related concepts, working of a debugger, using one of the most acclaimed debuggers i.e. gdb in different ways and scenarios, and various debugger-less ways of debugging.

Non-functional aspects of a program are as important and unlike in college, are given much attention in a corporate world. This chapter focuses on analyzing the program from a non-functional point of view, proper memory usage, performance, profiling, and coverage analysis of program. Also, the usage of specialized tools to fare the software against the real-world scenarios is described.

Quality analysis or Software Testing is a significant part of software development life cycle and there is lot more to it than just manually running the program and noting the success status. This chapter explains the importance, various phases, professional tools used, different methods of analyzing and testing a product, various types of testing and how it is done professionally in corporate world.

Chapter 8 Version Control System
Just think how important is the stage-save feature in a video game and you would realize why something similar is imperative in software development. If you don't get it, go through the chapter. ☺ Here you'll go through something that is of high importance but hardly taught about in any college – the basics of version control, its importance, usage and lingo is described to make things convenient for you from day one.

Chapter 9 Defect Tracking
The second last chapter of the book explains the defect tracking system and why choose it over an excel sheet to keep track of various issues and features related to the software. The underlying components, various stages a defect goes through, and integration of this system with Project management and Test Tracking systems are covered.

Chapter 10 Product Shipment cycle
Overall software development process is tightly coupled to Product Shipment Cycle and every software engineer needs to know and understand its details. As these things are not explained in colleges, rookies are often oblivious to various such concepts and take time to get hold of it. This final chapter explains these concepts like different kind of releases, release management, its flow cycle, and overall project management and lets one to be at comfort right from the beginning.

Appendix A
Error handling using Try-Catch block in Java

Appendix B
Installing open source software distribution

Appendix C
Integrating cscope with vim

Writing Good Code

"Hello World!!"

That's how your first piece of running code would have exclaimed as you started your journey into the world of programming. Yes, you have come a long way since and this chapter is going to help you further.

Writing Code – obviously that's more or less what you do as a professional programmer and get paid for! You already are capable of writing code to address a problem or implement an algorithm. This chapter is going to suggest how you can do it better and in more professional manner. Can you answer this simple question before proceeding –

Would you feel delighted when seasoned programmers shall appreciate your ability to write code proficiently right from the beginning of your career?
Yeah? Well, you're on the right track, continue.

In college, the evaluator and that's why you would primarily be concerned only about getting correct output. Aspects like Time and Space complexities, Hardware specifications, Portability across platforms, Performance and so on would be overlooked or seldom taken seriously. However, the scenario changes when you are working in a reputed organization where your code impacts numerous customers or users across the globe. The solutions you contribute to must better or at least be at-par with what competitors have to offer.

Enough said. Let's head to the business starting with discussion about a good code editor, followed by useful coding practices and guidelines while touching upon the trivial mistakes (sometimes blunders) committed by programmers. It's a promise to keep things crisp and not kill you with the extra-detailed instructions.

🔊 Coding in professional life is altogether a different ball-game. Better be prepared for it.

Code and Text Editor

You just can't write code in the air. An editor is required for programming and for professionals a good editor is a must.

Getting used to a new editor may seem arduous initially. However, once you get comfortable (which doesn't take long with little interest and persistence), it becomes an extension of yours and you'll be able to do all the necessary stuff as effortlessly as possible.

In case you wonder What-is-wrong-with-the-good-old-white-notepad? Well, nothing more than the fact that using notepad for programming in a professional environment is like using spoon for a shovel. While working on a project in any corporate environment, you need much more functionality than the basic Type-Cut-Copy-Paste. Also, browsing through the code across multiple files is certainly cumbersome with a primitive text editor.

Using a good editor enhances your productivity multi-folds. Take navigation for instance, pressing arrow keys repeatedly is certainly not the way a professional programmer traverses through the code. In any professional editor, there are shortcuts to navigate by characters, words, lines, code blocks and files so that you can accomplish more in lesser key strokes and without even touching the mouse. You'll soon realize that being able to scroll through the code and do operations like select, copy, cut, paste without a mouse gets work done much faster and efficiently. Also, working with multiple files within a single instance of the application makes it easier to manage and faster to maneuver.

The main characteristics of a good professional code editor can be summarized as follows:

Syntax Highlighting

This feature gives different color and font to source code as per the category of keywords and programming languages. It's quite helpful in catching common typos and leads to a better visualization and thus better understanding of the source code.

Auto-completion

Autocomplete feature gives a list of probable words even before you type them completely. Not only this means that you have to type lesser, but you also avoid unnecessary compilation error simply because you typed pTempArry instead of pTempArray.

Auto Indentation

Proper indented code is not only eye-catching but also aids in quickly understanding the flow of the program. It also helps to follow a common indentation pattern when multiple people work on same project.

Code Template / Skeleton

Some editors load a basic Template automatically as soon as you create a new file based on the programming language. It's not something without which you will feel incomplete but it is simply there to relax your typing fingers a bit.

Extensible and Expandable

Editor needs to be extensible and expandable so that the third party add-ons and plugins can easily be integrated. Typical examples would be the support for new programming languages and integration of tools like Cscope, perforce, etags, ability to run scripts, add macros and so on.

Configurable

What good an editor is if it is adamant and imposes its settings on you? A good editor is always configurable as per the user's needs and taste.

Code browsing

Code browsing is an integral part of an editor. Good editor provides advanced browsing support like multiple and split windows, browsing across the source code files and directories with easy to remember keyboard shortcuts.

Help

Most popular editors are the ones with the amazing in-built help and reference guides. The help for the command sets, keyboard shortcuts, and cheat sheets should be readily available in case you need them.

Quick Navigation

Editors like vi and emacs are the most preferred and popular choice of the professionals mainly because of the multiple and easy to use navigation options available to them. You can navigate across the code while skipping word, lines, block of statements, files, even directories.

Availability on all platforms

Be sure to check the availability of your choice of editor on different platforms. This is really important because once you get the knack of the editor and start getting addicted to it, you don't want to learn a new editor if you have to work on a different platform.

◀)) Get familiar with at least one good code editor.

Coding Guidelines

Every development team has its own set of rules and standards which are devised over a period of time. Being a new member to the coding fraternity bestows those coding practices on you. Having a good understanding of the coding practices followed in your team helps you in multiple ways:

- Easier grasp of existing code
- Avoid scornful looks of some programmers who are really choosy on following coding guidelines
- Coding guidelines are developed after a lot of discussion by experienced programmers and hence can benefit you in writing better and optimized code
- Following the same coding guidelines leads to better, faster and stream-lined code reviews

Some commonly used practices that are used as coding guidelines in many companies are listed below. These are no way the only coding standards that you'll need to follow, but will get you more than prepared for the professional world.

Variable naming convention

You would agree that the names of the variable should reflect their usage. But, there is one more attribute that may be attached to the variable - its data type. As soon as you start browsing through your code base, you may find funny names like - pch, pvfreeMemory. Although, these names seem fuzzy and hard to pronounce but only if you haven't heard about the "Hungarian" notation that was developed by Charles Simonyi. It uses proper capitalization to distinguish between separate words. An alternative to this could be using "_" (underscore) to separate words in a variable name. Let us have a look at some examples:

```
1.  char ch;
2.  byte b;
3.  flag f;
4.  char *pch; /* p -> pointer */
5.  char **ppch; /* pp -> pointer to a pointer */
6.  static int sitemp; /* s -> static, i -> integer */
7.  int gcount; /* g -> global */
8.  typedef int tLength; /* t -> typedef */
9.  int aList[MAX_LEN]; /* a -> array */
10. void *pvfreeMemory(void *pv);/* pv -> function
    returning a void pointer */
```

Indentation

You would have been using your set of rules for indentation and most certainly that would be to recognize or differentiate your pattern and code amongst others. Works for a loaner but in a team all should follow a common standard even for indentation. Embrace that with a smile. Here are some most commonly followed practices:

```
1.  int imyfunc(void *pv)
2.  {
3.      int i;
4.      char ch;
5.
6.      if (pv == NULL) {
7.          /* statement 1 */
8.          /* statement 2 */
9.      } else {
10.         /* statement 3 */
11.     }
12.     i = itempFunction(pv, ch);
13.     return (i + 5);
14. }
```

- Always use tabs instead of spaces for indentation at the beginning of a line.
- Avoid inconsistent use of tabs and space. It results in showing unwanted differences when comparing files.
- Indent a statement block at same level (Statement 1 and 2 above).
- Provide a space between the first parentheses after a keyword. For example, Check the space between "if" and the next "(". Also, note there is no space between the function name - itempFunction and the "(" following it.
- Provide spaces around operators in an expression (spaces around "+" operator in return statement)
- Separate parameter list of a function call with space after comma (check the function call to ctempFunction)
- Let the length of your line not exceed 80 chars.
- If the length of a line is going to exceed 80 char limit, then either break after comma, or before an operator and align new line at

the same level where the expression started on the previous line. Getting confused? Following examples should help:

```
1.  longVariable = veryLongFunction (aLongVariableName1,
2.                                   aLongVariableName2);
3.  longVariable = veryLongFunction(longVariableName1
4.                                  || longVariableName2);
```

Declaration

Have a look at the following declaration. Do you find anything amiss?

```
1.  int* pch1, pch2;
```

What does first glance at this statement looks like - declaring two integer pointer variables? What it actually does is that it declares an int pointer variable and an int variable. One may end up assigning a pointer to int which may result in some issue. The compiler would throw a warning in such cases but it's better to avoid such easy-to-overlook mistakes by just following a strict rule of "One declaration per line". Check the declaration below and then decide which one is easy to understand and unearth silly mistakes.

```
1.  int* pch1;
2.  int pch2;
```

The declarations should be the first thing to write at the beginning of a block. It is better to separate the declaration and initialization of variables in a block. Not only it improves the aesthetics of the code but helps in catching un-initialized variables easily.

Have a look at the code below. It seems quite evident that the initialization for 'flag' was either left out intentionally or someone missed it.

```
1.  int itempFunc(void)
2.  {
3.      // variable declarations
4.      int i;
5.      int flag;
6.      char *pch;
7.
8.      // variable initialization:
9.      i = 0;
10.     pch = NULL;
11.     /* ---- */
12. }
```

While declaring variable names it's advisable to avoid using the same variable name in inner blocks of code. Not only it makes the intentions of the programmer unclear, but also becomes a pain to debug that piece of code.

It is a good coding practice to provide a function header clearly stating its input and output parameters, a short summary of the coding algorithm and the expected function call syntax.

```
1.  /*
2.   * ifree_list() first frees up the list container
3.   * along with the
4.   * list items. If succeeds, then sets the list pointer
5.   * explicitly to NULL.
6.   *
7.   * pplistptr -> pointer to a pointer to type tlist.
8.   * retuns the status TRUE if succeeds, else returs
9.   * FALSE
10.  * and leaves
11.  * the pointer plistptr unmodified.
12.  *
13.  * int iStatus = ifree_list(&plist);
14.  */
15. int ifree_list(tlist *pplistptr)
16. {
17.     /* -- */
18. }
```

Statements

Just like declarations, follow the rule of "One Statement per Line" for statements also. Do not write overly complex code by stuffing in multiple coding statements in a single line. It makes the code un-readable for other developers. Also, enclose compound statements in curly braces even if there is one single statement. Many times this simple coding practice avoids hard-to-find bugs.

Comments

Comments are as important as the code lying underneath them. It is considered a good programming practice to include comments in your code so that you are not only helping fellow developers, but yourself too if you plan to revisit your own code. Believe it or not, after a year even you yourself would forget why you introduced that certain if condition.

- Code and the comments that describe it should be indented at the same level.

```
1.  if (pch == NULL) {
2.      /*
3.       * This is a multi - line comment
4.       * pch is NULL. Dont proceed forward and return
5.       * the proper
6.       * Error code from here.
7.       */
8.      return ERR_INVALIDVAL;
9.  } else {
10.     // This is a single line comment
11.     pch->errbits = 0 // A trailing comment
12. }
```

- Explicitly comment the Follow-through condition in a switch - case statement.

```
1.  case ADD:
2.      /* Follow - Through */
3.  case SUB:
4.      handleBinaryOp();
5.      break;
```

🔊 Following coding guidelines saves you a lot of hassles while writing code and code review. Stick to these throughout your life.

Useful Tips and Common coding mistakes
Hope you had fun with the coding guidelines!

This section shall provide some useful Tips to help you write better code while also describing some common coding mistakes that programmers make. The following is by no means an exhaustive or complete list but simply a self-compiled one which should get you going on the freeway of writing better code. After all, isn't that what we want?

Use Conditional macros to insert and/or clip out code.
These are used in the following manner:

```
1.  #ifdef <token>
2.  /* code */
3.  #else
4.  #define <token>
5.  /* code to include if the token is not defined */
6.  #endif
```

#ifdef checks whether the "token" has been #defined earlier in the file or in an included file. If Yes, only the code between #ifdef and #else is included. Otherwise, the code between #else and #endif is included. These are used generally to create compilation switches based on certain flag(s).

Here are some common mistakes made by developers that can be eliminated by using conditional macros:

- ***Not removing the debug logs from non-debug builds***
  Logs are a necessity as they let the user know if everything is going well and if not, where can be the issue. However, writing to a file very much affects the performance of the software. Therefore, the logs which are needed only for deep debugging should not go into production builds.

One of the authors could make a perceptible improvement to the performance just by eliminating unnecessary debug logs from production builds using conditional macros. With these, all you need to do is define/modify a token DEBUG_LOGS in the Makefile and define log function as follows:

```
1.  #ifdef DEBUG_LOGS
2.  #define LOG(args) fprintf(args)
3.  #else
4.  #define LOG(args) ((void) 0)
5.  #endif
```

If DEBUG_LOGS is defined in Makefile, LOG(args) shall translate into log statements. Otherwise, LOG(args) statements shall become void.

- **Multiple inclusions**

  You know that #include directive directs the preprocessor to treat the specified file as if its contents had appeared in the source program at the point where the directive is. It is common to #include a header file in another header file and therefore multiple inclusions may occur unintentionally. For example,

```
File "grandpa.h"
struct somestruct {
    ...
};
```

```
File "pa.h"
#include<grandpa.h>
/* Some definitions */

File "kid.c"
#include <grandpa.h>
#include <pa.h>
/* Some code*/
```

As the file kid.c includes multiple copies of the content of file grandpa.h, the compiler shall issue an error. However, including standard C Library files any number of times doesn't show any error. Why? Because the body of these header files is always enclosed in #ifndef...#endif block. This is termed as *Include Guard*:

```
1.  #ifndef <token>
2.  #define <token>
3.  /* code */
4.  #endif
```

Guard:
The first time inclusion of code defines the token which doesn't allow further inclusions. In the above example, grandpa.h can be redefined as follows to avoid this issue:

```
File "grandpa.h"
#ifndef GRANDPA_H
#define GRANDPA_H
struct somestruct {
    ....
};
#endif
```

Note: Instead of **#ifdef** and **#ifndef**, **#if defined ()** and **#if !defined()** can also be used respectively. The advantage is that the latter allow for checking multiple and complex conditions as well. For example,

```
1.  #if defined(x) || defined(y) && !defined(z)
2.  ...
3.  #endif
```

Proper use of typedef keyword

The purpose of *typedef* keyword is to assign alternative names to existing data types, often to those whose standard declaration is confusing or likely to vary across implementations.

Therefore, it helps not only in indicating more closely what the variable indicates but also in defining platform independent datatypes. For example, we can define a floating point type called MYTYPE that has the highest precision available on the machine:

```
1.  typedef long double MYTYPE;
```

On machines that don't support long double, this typedef will look like this:

```
1.  typedef double MYTYPE;
```

And on machines that don't even support double:

```
1.  typedef float MYTYPE;
```

Note: In some books/forums/blogs, it can be found mentioned that typedef defines a new type. However, according to "The C Programming Language" by Kernighan & Richie:

"It must be emphasized that a typedef declaration does not create a new type in any sense; it merely adds a new name for some existing type. Nor are there any new semantics: variables declared this way have exactly the same properties as variables whose declarations are spelled out explicitly."

Here is a common typedef related mistake made by developers when they try to use typedef and #define interchangeably:

At first, #define and typedef may seem similar but there is a subtle difference: *#define* are just replacements done by preprocessor but *typedef* are handled by the compiler itself. In essence, a *typedef* is similar to a *#define*, except that since it is interpreted by the compiler, it can cope with the textual substitutions that are beyond the capabilities of the preprocessor. Consider the following example:

```
1.  #define STRING char*
2.  typedef char* String_t;
3.  STRING str1, str2;   /* This has a subtle issue */
4.  String_t str3, str4; /* This does not */
```

Here, **str1**, **str3** and **str4** get declared as char pointers but **str2** gets declared as **char**, which is probably not the intention. Perhaps substituting the #define would make it more clear:

```
1.  char* str1, str2; /* Only the first value gets declared
    as a pointer */
```

Keep it static

static simply means "unchanging" and has seemingly unrelated uses. *static* is used in following two ways:

- **Access controller**
  A global variable or a function defined as static is restricted to the file where it is defined. This limits the visibility of the variables/functions and thus avoids naming conflicts. This usage of static keyword is more popular than the other one.

- **Value Retainer**
  A local variable defined as static inside a function retains its value between invocations. For example:

```
1.  int function()
2.  {
3.      static int a;
4.      return ++a;
5.  }
```

shall return the values 1, 2, 3 ... on subsequent calls. This is generally used in coding of some embedded systems like ATM machines.

The RETURN of....

As you know, a return statement simply ends the processing of the current function and returns the control to its caller. There is no limitation on number of return statements that can be used. However, it is quite common to find clumsy and erroneous code when multiple return statements are used. The common blunders made are:

- ***Not concluding the commenced***
  It is imperative that you conclude what was commenced in a function. For example, leaving a file opened or not freeing the memory that is not needed anymore are the common but dangerous mistakes.

  In functions having multiple return statements, it is quite usual to find such mistakes because in large projects, the person who modifies the code is not the one who has written it. A new person adding another condition for return may miss taking care of everything.

- ***Bulky and repetitive code***
  Adding a log, file close and freeing memory before each return statement simply adds to the size of code and makes it look clumsy. To avoid such issues, use minimum return statements in a function unless unavoidable. Generally, it is considered better to have a single point of return in a function also termed as "The single-return law". Such practice reduces the chances of aforementioned mistakes, helps in debugging and in most of the cases improves the readability of the code.

```
1.  int someFunction(char *str) {
2.      FILE *fp;
3.      char *tempStr;
4.      if (!str)
5.          return -1;
6.      // Some code to allocate memory to tempStr
7.      if ( someCondition ) {
8.          free(tempStr);
9.          return 1;
10.     } else {
11.         // Code to open a file and do an operation
12.         if ( someOtherCondition ) {
13.             free (tempStr);
14.             fclose(fp);
15.             return 2;
16.         } else {
17.             // Some more processing
18.         }
19.     }
20.     if (tempStr) free(tempStr);
21.     if (fp) fclose(fp);
22.     return 0;
23. }
```

Instead of writing code like this, it could have been written better as:

```
1.  int someFunction(char *str) {
2.      FILE *fp;
3.      char *tempStr;
4.      int retVal = 0;
5.      if (str) {
6.          // Some code to allocate memory to tempStr
7.          if ( someCondition ) {
8.              retVal = 1;
9.          } else {
10.             // Code to open a file and do an operation
11.             if ( someOtherCondition ) {
12.                 retVal = 2;
13.             } else {
14.                 // Some more processing
15.             }
16.         }
17.         if (tempStr) free(tempStr);
18.         if (fp) fclose (fp);
19.     }
20.     return retVal;
21. }
```

However, in case of nested conditional statements i.e. the function has a conditional behavior that does not make clear what the normal path of execution is, usage of Guard Clauses is preferable. For example,

```
1.  int function() {
2.      int result;
3.      if (condition1)
4.          result = function1();
5.      else {
6.          if (condition2)
7.              result = function2();
8.          else {
9.              if (conditon3)
10.                 result = function3();
11.             else
12.                 result = function4();
13.         }
14.     }
15.     return result;
16. }
```

Can be better written as:

```
1.  int function() {
2.      if (condition1) return function1();
3.      if (condition2) return function2();
4.      if (condition3) return function3();
5.      if (condition4) return function4();
6.  }
```

Another neater way is to implement the Java like Try-Catch blocks in C. See Appendix A for an example of Try-Catch block in Java.

Proper use of Brackets

Brackets make for an essential part of any programming language including C. The C pattern of parentheses usage has prevailed in most of the modern programming languages as well:
( ) Parentheses for functions and expressions
[ ] Square brackets for arrays
{ } Curly brackets or code blocks

The parentheses hold the highest precedence in C and their proper usage can avoid many common mistakes made by the programmers and make code more readable. Some examples:

- ***Mixing different kind of operators***
  The following expression is expected to stuff two 8-bit bytes into a 16-bit word.

```
1.  Word = ByteA<<8 + ByteB;
```

  Does it? No, because + operator has higher precedence over the shift operator. Result – shift ByteA by 8+ByteB.

  The best practice is not to mix up different type of operators. Programmers don't seem to have much trouble remembering precedence order in same class of operators but need a reference while dealing with different kind of operators. It would be better to either write the above mentioned expression as

```
1.  Word = ByteA<<16 | ByteB // using Bitwise operators
```

  or

```
1.  Word = ByteA*256 + ByteB // using Arithmatic operators
```

  Or else, just use parentheses to avoid the confusion and error:

```
1.  Word = (ByteA<<8) + ByteB
```

- ***Using assignment and equality operators in a single statement***
  Consider the following pieces of code which aim to open a file in read mode and check if it could be opened:

```
1.  if (NULL != fp = fopen ("filename", "r")) {
2.  ...
3.  }
4.  ----------------------------------------------------
5.  if (NULL != (fp = fopen ("filename", "r"))) {
6.  ...
7.  }
```

Which one would you prefer? The latter is not only correct but is better in terms of reading and understanding.

Again, the best practice would be not to mix different type of operators and use them separately as shown below:

```
1.  fp = fopen("filename", "r");
2.  if (fp != NULL) {
3.  ...
4.  }
```

However, many programmers tend not to follow this. Therefore, they must use parentheses liberally to avoid issues.

Be Assertive

"This memory allocation cannot fail."
"The count can never be negative."
"This parameter can never become NULL."

One of the biggest mistakes programmers make is to assume. As the saying goes, ASSUME simply means making ASS out of U and ME.

If you think something can't happen, make sure it won't! One of the best aids provided by C for the purpose is assert() or _assert() macro that checks for a boolean condition. It is defined in ASSERT.H, and has prototype:

```
1.  void assert(int expression);
```

If the expression evaluates to TRUE, assert() does nothing. If expression evaluates to FALSE, assert() displays an error message on *stderr* and aborts program execution.

If a pointer passed to your function should never be NULL, do an assert check for it as:

```
1.  void someFunction(char *string) {
2.      assert(string != NULL);
3.      ...
4.  }
```

This assert statement can also be written as *assert(string)* which gets translated to *assert(string!=NULL)*. *assert* is of great help but you need to be careful in order to avoid common mistakes that programmers make while using it.

- Don't pass an expression to assert that may cause a side effect. assert (string = NULL) shall cause string to be assigned NULL posing issues in further code.
- Putting the "must be executed" code in assert - Assertions can be turned off at compile time. Therefore, never put the code that must be executed into an assert.
- Replace real error handling with assertions – As mentioned above, assertions are to document the logically impossible situations and identify absurdities in programming – If the impossible occurs, something is radically wrong. This is different from error handling: most error conditions are possible, perhaps highly unlikely to occur in practice but possible. For example, writing code like this is a bad idea:

```
1.  printf ("Enter 0 or 1 : ");
2.  scanf ("%d", &i);
3.  assert ((i == 0) || (i == 1)); // Bad usage of assert
```

Judiciously Allocate and Free the memory
malloc() and free() are respectively used for dynamically allocating and freeing the memory chunks in C. The following section explains

how these can be better used to avoid the goof-ups generally made by programmers.

- **Memory leaks**

  If the dynamically allocated memory is not freed after its not required anymore, it results in memory leaks causing program size to continue to increase and finally crash. A programmer must make sure to free the dynamically allocated memory after its purpose is served.

- **Returning the reference to a local variable**

  Can you identify the issue with the following code?

```
1.  int * someFunction() {
2.        int i;
3.        i = 8;
4.        return &i;
5.  }
```

This function returns the reference to a local variable. The memory allocated for the local variables is automatically released as soon as the function processing is complete and control is returned to the caller. So, returning pointer to some memory which is no longer valid results in undefined behavior.

To return the reference and avoid undefined behavior, either the variable must be declared static or must be allocated dynamically. In these cases, the allocation is done on heap and it won't be released by the compiler. This shall work:

```
1.  int * someFunction() {
2.        int *i;
3.        i = (int*)malloc(sizeof(int));
4.        *i = 8;
5.        return i;
6.  }
```

Note: Programmer needs to take care of freeing the dynamically allocated memory when it is no longer required otherwise it shall cause a memory leak.

- **Not setting the pointer to NULL after freeing it**

  In C, there is no way to check if a pointer is actually pointing to a valid location. Even after freeing the memory, the pointer shall have reference to the memory location unless set to NULL. It leads to the following common issues:

  Double freeing

  Trying to free an already freed memory that shall lead to a crash.

  Reference freed memory location

  By the time of referencing freed location, it may get allocated for some other purpose. Referencing/Writing to it may cause data corruption and undefined behavior.

  Setting pointer to NULL after freeing is a defensive practice and protects against above mentioned issues. You may define and use the following FREE macro to take care of mentioned points:

  ```
  1.  #define FREE(x) if(x != NULL) { free(x); x = NULL; }
  ```

- **Accessing what is not yours**

  Issues arise when you try accessing the memory what is not allocated to you. *Array index out of bound* error is an example you would have come across. Not initializing the pointer(s) to NULL leads to similar issue. Observe the following code-

  ```
  1.  void someFunction() {
  2.      char *ptr;
  3.      if (ptr) {
  4.          free(ptr);
  5.      }
  6.  }
  ```

  The program shall crash with Segmentation Fault because ptr would be pointing to some random location in memory. Just initializing *ptr* to NULL would avert the issue.

  Note: Rule of thumb - initialize all the pointers to NULL.

This list is never going to end but now you know a good amount of what you need to about Writing Good Code. Let's stop here to give you and ourselves a break. In the meantime you may go ahead and play around with your code and punctuate it with the tips learned throughout this chapter. Try to make it more professional both logically and aesthetically. Let's meet in the next chapter with detailed information on the way we browse and navigate through a code base.

Chapter Review

- Code base and coding environment is totally different in professional environment. Not only does it require special skill-set but also special tools and utilities.
- The main programming concerns you should be concerned about are coding complexity, scalability, compiler dependency, code readability, Code review, Performance and conventions followed while writing code.
- You need much more than a basic cut-copy-paste editor, if you want to be an efficient coder. You need to learn at least one advanced editor.
- An Advanced editor like vim, Emacs has features like – Syntax highlighting, Auto-completion, auto indentation, extensibility, configurable, built in help, Quick navigation. Such an editor is available across all the platforms and is expandable to third party plugins.
- Coding guidelines are the standards and rules laid down by experienced programmers of the corresponding team. You need to quickly adapt and follow them.
- Coding Guidelines helps in easily understanding the already written code. You develop the skill of writing better and optimized code and in the process you gain confidence of fellow programming fraternity.
- There are a lot of common programming mistakes which can be avoided by following careful programming practices. If left untended, they might lead to potential bugs.

2

Code Browsing

Wondering if you should skip this chapter? After all, what can possibly be there in browsing the code that qualifies to consume a whole chapter? Read on to find out, you wouldn't be disappointed.

Till now you would have worked on mini projects or stand-alone programs. The characteristics of such endeavors would be: consisting of few files and very few contributors making modifications or adding new functionalities. Due to small size of code, you might have known your code like the back of your hand. Basic search functionality would have sufficed if you ever had to browse through code. However, in real world you shall find plethora of code even for small projects with hundreds of people contributing round the clock to its development. You may download and see the source code of any open source project like VLC Media Player, Firefox or Chrome to get an idea.

It may appease you that becoming proficient in code browsing in far easier than any other aspect related to programming. Just with little practice you shall be browsing through any size of code; as effortlessly as a pro.

Let us start by listing down the features that are worth learning as you find yourself evolving into a professional code browser:

- Quick Navigation
- Multiple Files in a single view
- File and Directory view
- Browsing History
- Integrated pattern search in files and directories
- Functions callers and reference listing
- Ability to list header file inclusions
- Token Search specific to programming language
- Hyperlink Navigation
- Execute commands

🔊 A good code browser lets you remain focused on browsing by taking care of your distractions by providing integrated features and other tools.

In this chapter you shall be briefed about various tools available for the purpose. You will require multiple tools in order to meet all of the above mentioned goals. First choose a good code browser that meets most of the conditions laid down before you. Editors like Vim, Emacs, Source Insight offer cool keyboard shortcuts to achieve near perfect navigation, multiple views in one single pane, integrated directory and file browsing. They store and maintain browsing history and have the ability to search keywords or patterns within multiple files. Vim and Emacs have the added benefit of allowing one to execute commands from within the browser.

Although good editors provide a way to perform search over the code spread across multiple files, but it is not an optimal solution for most of the scenarios that you shall be handling on a daily basis. You need specialized tools to efficiently perform the search operations. Cscope is one such open source tool that gives us customized search options specific to the programming language.

The best way to learn anything is by doing it. It is actually easier done than said (or Read in this case). It is really easy to set up the environment if you want to learn code browsing by practicing along with reading. There are 3 simple steps to follow that should get you going:

* Download the code: Download a sample source code of any open source application available online.
* Download Your Tools: You can easily download the code browsing tools that we are going to explain from their respective webpages. Cscope, ctags, etags, vim, emacs all are easily available for almost all of the flavors of the present day Operating Systems and hardware.
* Setup your tools: Install the tools from the source code you just installed. It's really easy to do that if you follow the instruction given on the homepage of each of the application you want to install. Once installed you are ready to use them as standalone or integrated with other tools that we will explain as and when needed.

Detailed instruction can be found in appendix B.

Searching the function definition using Cscope is definitely time-saving but still it can't beat the hyperlink navigation. Tags and Cscope when integrated with code editors/browsers helps us to directly jump to the required code piece with a simple click of mouse or a keyboard shortcut. This is analogous to clicking a hyperlink on a webpage.

The rest of the chapter revolves around all of the tools and utilities discussed above with specific examples. Let's start with the introduction to Cscope, its usage and integration with code browsers. After Cscope, we will uncover Tags utility and finally we will discuss in detail about the code browsers.

◀ঠ Efficient code browsing is a daunting task. You will need multiple tools and their integration with each other to help you coast along.

Why Cscope?

Remember, we listed down few important features that are a must for a good code browser tool. Following is an elaboration of few of these aspects in detail with reference to Cscope.

Token Search specific to programming language

With the increasing number of source files and the size of individual files it is advisable to have a tool that does the job efficiently. Tools like Cscope create their own customized search database according to the programming language of your files. It is highly efficient and stable for keyword and token search for that specific programming language as compared to other tools.

Functions calling a specific function

Using simple pattern search for this query gives hundreds of redundant result which you might not be interested in. Cscope is one such readymade tool specialized to return only relevant functions calling the function provided as search token.

Function calls made from a function

If you were to list down all the functions that are called from within a function – func1(). Easy, isn't it? Just open the file containing the function and find the required info. But, what if the func1() itself is some 2000 lines long? Would you still go for the manual method you used earlier or seek help from a specialized tool like Cscope that can perform the task with just a few simple keystrokes? Choice is yours!

Ability to list header file inclusions

This is more or less similar to whatever was just discussed regarding function calls. It is very important and time saving if you have a tool at your disposal that can list down all the .c files that include a specific .h file.

Cscope is a command based graphical interface that allows easy and efficient code browsing for the developers. It was originally developed by Joe Steffen at the Bell Labs and is currently available freely under a BSD License. It has been used to manage projects involving 20 millions of code till now! That should give you an idea about the acceptance of Cscope by the developer's community around the world.

Generate cscope.files

To get started, you need to initialize Cscope database by generating a file named cscope.files. This file is going to have the listing of all the source files that you wish to search in future. Use the 'find' command present on unix machine to achieve the same:

```
1.  $ cd myproject
2.  $ find `pwd` \( -name "*.c" -or -name "*.h"⊠ \) >
    cscope.files
```

'find' command is used to list down all the ".c" and ".h" files in the current directory ("pwd" command gives the current directory name). Finally '>' operator is used to redirect the list of files into a new file "cscope.files". Simple, isn't it!

In some flavors of Unix and Linux you can create the Cscope database by simply executing command "cscope –R" in the parent directory under which your source code exists. This command will generate the list of file recursively by iterating through all the sub-directories. By default, it includes file with .c, .h, .i and .y extensions.

Generating Cscope database

```
1.  $ cd myproject
2.  $ cscope -b -k
```

'-b' option tells Cscope to build database while '-k' option sets its kernel mode. With kernel mode set, Cscope searches for C library source files in the included Cscope files only and not in the /usr/include/ directory. This command will generate a file named 'cscope. out' in the same directory which is nothing but the database which you were trying to build.

Using the database

To use Cscope database just generated, simply run the following command:

```
1.  $ Cscope -d
```

'-d' option tells Cscope NOT to re-generate the database. If you ignore this option from the above command, Cscope will recreate its database from scratch.

Have a look at the options displayed at the bottom of the cursor based text screen when you execute 'cscope –d' on your shell.

```
Find this C symbol:
Find this global definition:
Find functions called by this function:
Find functions calling this function:
Find this text string:
Change this text string:
Find this egrep pattern:
Find this file:
Find files #including this file:
```

You can use the combination of <TAB> and arrow keys to navigate around this menu and provide your keyword to search. Let us try it out with our sample source code of vim editor, details on how to download source code can be found in Appendix B. Suppose you wish to find out all the callers of the function "find_help_tags()". Here is what you will get by pressing Enter after typing "find_help_tags" in the query "Find functions calling this function:"

```
Functions calling this function: find_help_tags

File            Function           Line
0 ex_cmds.c   do_help            3852 n = find_help_tags
                                       (arg, &num_matches,
                                       &matches);
1 ex_getln.c  ExpandFromContext  2794 return find_help_tags(pat,
                                       num_file, file);



Find this C symbol:
Find this global definition:
Find functions called by this function:
Find functions calling this function:
Find this text string:
Change this text string:
Find this egrep pattern:
Find this file:
Find files #including this file:
```

Although the output is self-explanatory, here is the explanation for the sake of completeness:

Function "find_help_tags" is being called at two other places and the Cscope output points out the file name and the exact line number with the code snippet too! And this is not all it has to offer, you can use arrow keys to scroll over the results displayed and press <ENTER> key to view the full code in vi editor (you can change the editor by overriding the environment variable EDITOR). Press ':q' in the vi and you will be returned back to the same Cscope result window and you can scroll around to check another result. Press Ctrl+D to quit Cscope.

That is almost everything that you need to know about Cscope if you want to use it in as a standalone tool. For further studies you can refer to the man page of Cscope here:
http://cscope.sourceforge.net/cscope_man_page.html.

Cscope is a wonderful tool as it provides freedom of choice. It can easily be integrated with vim and that makes it a wonderful tool to use. You can simply search for required stuff from the same editor window that you use to write code. Some developers prefer using it as a standalone tool, while most like to have it integrated with the editor of their choice. We will explore this integration at a later stage during the discussion of vim editor in detail.

🔊 While you are working on a project where your code base is regularly changing, you need to update your Cscope database periodically.

TAGS

By definition, a tag means a reference, a pointer or a bookmark to a given object. Tags utility brings the same concept to the programming jargon. In a program, tags may refer to a function, structure definition, data type, macro, subroutines. The idea is to assign a tag to each of them so that we can perform a highly efficient search over them whenever needed.

The exact line number and filename in a C file can act as a unique tag for the definition of that function. The same concept can be extended to other objects to build up a tags file which can be referred by the tags utility to search a given token. Thus, generically speaking a tags file is quite similar to an index file of the code base that helps to find the definition of its elements in an easy and quick manner.

In Unix based machines, tags itself comes into several different flavors. Inherently they serve the same purpose and possess the same structure, but they differ in the way they can be integrated with code editors. The most popular one is called *ctags*. The original *ctags* was introduced by Ken Arnold for the BSD Unix system. *Ctags* are easily integrated with several text and code editors. Relationship between *ctags* and vim editor is quite old and it is the most used tag utility. *Etags* is the *ctags* utility that comes with another popular text editor – Emacs.

There are various other language specific tag utilities supported by specific text editors. "*Hasktags*" is the tag utility for Haskell source files, while for javascript we have "*jsctags*" and so on.

Similar to the Cscope utility, tags needs to create an internal database which can be referenced later for search queries. Tags file discussed previously serves the same purpose. Let us have a look at some examples to have a practical idea about how a tag file looks like and more importantly how to generate one.

Creating tag file

Different tag utilities require different command line options to generate the tag file and we are not going to get ourselves bored up with learning all of them. Let us target only the most popular ones.

ctags is specific to C and C++. The format and language of the source files is internally learnt by the tool from the file name and the file extension. Executing the command "*ctags **" will generate the tag file from all the recognized source files in the present directory. However, "*ctags –R*" will generate the tag file in the current directory but will recursively take all the files and directories into consideration. You can also use the *cscope.files* generated in previous section to selectively include source files and pass it as an argument to the command ctags:

```
1.  $ cat Cscope.files | ctags -
```

For detailed options you may wish to refer to the official man page of the ctags here:
http://ctags.sourceforge.net/ctags.html

If you don't want to look for *cscope.files*, use this command to generate the etags recursively covering all the ".c" and ".h" code files that lie under the current directory and subdirectories.

```
1.  $ find . -name "*.[ch]" -print | etags -
```

Etags users may execute "etags –help" to refer to the several command line options available for the *etags* command.

Understand the tag file

ctags appropriately names the generated tag file as "tags", while etags names the file as "TAGS". The following entry is taken from the tags file generated using ctags utility for the same sample source code that we used for Cscope.

```
1.  mch_getenv /var/temp123/vim-5.8/src/os_unix.h /^#define
    mch_getenv(x) (char_u *)getenv((char *)(x)/
```

The above entry can be easily broken down as :
Tagname<TAB>file_name<TAB>ex_cmd

- Tagname refers to the identifier whose index is stored here
- <TAB> is exactly one Tab character
- file_name is the source file where the definition exists
- ex_cmd refers to the command set that can be executed by editors like vim to jump to the definition of the tag. It could be as simple as a search pattern or line number. (As of now you can leave the details, it will become clear when we touch upon the integration of ctags with text editors).

The "TAGS" file generated by etags consists of several sections – one section per source file. Non-Ascii characters are used to punctuate the section entries which are plain text otherwise. A file section in the tags file start with a header generally split in two different lines. It looks something like this in the sample TAGS file generated for our sample project:

```
1.  ^L
2.  /var/temp123/vim-5.8/src/buffer.c,1429
```

The first line consists of special character with value "oc" in hexadecimal (hereafter referred as oxoc), while the syntax for the second line has the following format:
file_name, <size of tag definition data in bytes>

Soon after the header, the tag definitions start with one definition per line. Following is a snippet from our sample TAGS file:

```
1.  static char_u *lasttitle ^?lasttitle^A1982,47245
```

The format being:
tag_definition<0x7f>tagname<0x01>linenumber, byte_offset

Here, phrase "tagname<0x01>" can be omitted if the tagname can easily be deduced from tag_definition.

If you notice, there is no file name in the tag definition mentioned above. That is simply because the definitions belonging to one source file are found under their corresponding section header. Hence, the filename needs to be deduced from the parent section header present in the tag file.

Using the tags file

Searching into a single file for a token is definitely much better then searching in the whole source tree. Once the tags file is generated in place, you can use it as an index file to quickly find out the basic elements / tokens of the source code you are using. But, this is not the end of the world when it comes to using tags file. The real use comes into picture when the tag files are integrated with your text editors and source code browsers. This subject will be a part of discussion at a later stage in this chapter in the context of hyperlink navigation from a code browser.

◀)) Database generated by the tags utility is static and hence needs constant update every time there are significant code changes.

Code Browsers

This section covers the aspects of code browsing from the viewpoint of code browsers. *Cscope* and *Tags* are specialized tools for performing a certain operation; they don't have the capability to satiate all our needs. We need a good browsing tool that has inherently almost all the capabilities built in itself and more importantly it has a cool user interface and easy to memorize short cuts.

Code browsers are abundantly available, whether they are open source or proprietary software. But, for the sake of brevity the ensuing discussion is restricted to the most popular and most efficient ones. If you get a chance to work on UNIX or its derivatives you are bound to cross paths with *Vim* and *Emacs*. On the other hand if you were to work on a Windows machine, then you will surely have people in your team working on *Source Insight* or an IDE. In short, all this fuss is created just to let you know that the focus here will only be on Vim, Emacs and Source Insight.

Quick Navigation

Navigation is the most essential component of code browsing. Quick navigation means the ability to move around the open file with minimal effort. Files having few thousand lines of C code are pretty common; just imagine if you were to move on to line number 5986 simply using down arrow key. Contrast this situation to the ease with which you can jump onto a specific line number by simply pressing ":5986" on command line in Vim editor.

If that example doesn't convince you, take another practical situation that occurs so often while browsing code. Suppose you are looking at a highly nested but badly indented for loop in a C file. How do you figure out where a particular closing brace is for the opening brace you have your cursor on. You got to either bang your head and find it manually or move your cursor on the opening brace and simply press "SHIFT+5" key inside Vim editor and the cursor will simply jump to the closing brace automatically. Phew! Wasn't that quick? Good code browsers have loads to offer, it is up to you how much to exploit them.

Navigation ability should not be there for the namesake purpose only. What marks a good code browser a good navigator too is its command set. The command set should a simple pattern of keystrokes for which you need not scratch your head every time trying to remember it or go through a boring manual of the tool.

Following is a list of some of the useful navigation operations and the shortcuts to achieve them:

Navigation operation	Vim	Emacs	Source Insight
Start of line	0	Ctrl+a	Home
End of line	$	Ctrl+e	End
Beginning of word	B	Alt+b	Ctrl+Left arrow
End of word	E	Alt+f	Ctrl+Right arrow
Page Up	Ctrl+b	Alt+v	Page Up
Page Down	Ctrl+f	Ctrl+v	Page Down
Start of File	Gg	Alt+[	Ctrl+Home
End of File	G	Alt+]	Ctrl+End
Start of parenthesis	%	Ctrl+Alt+b	
End of parenthesis	%	Ctrl+Alt+f	

Options listed above don't even make up to the least of what these powerful browsers have to offer. This is just to get you started. For further documentation you need to refer to their corresponding manuals:

Vim - http://vimdoc.sourceforge.net/

Emacs - http://www.gnu.org/software/emacs/manual/

Source Insight - www.**sourceinsight**.com/docs35/User**Manual**.pdf

Multiple Files in a single view

You are diligently writing some important piece of code and suddenly you feel the urge to have a quick look at some other function. For that you need to open that file in your code browser and most importantly shift your focus to a different window or tab. But, that's alright. Next you plan to refer to the newly opened file but still want to have a look at your own routine to tally few of the lines of the code. Don't you find it irritating when you have to shift your focus from one file to another and that too multiple number of times?

Well, there is an easy way in editors like Vim and Emacs where you can split your current window either horizontally or vertically and open a new file in half of the window, while having the liberty to have your older file still visible in other half. And the good news is that window splitting doesn't stop at only 2 windows in a single view, you can open as many windows as you can handle.

Listed below are few of the shortcuts you will require to work through multiple files in single view:

Navigation operation	Vim	Emacs
Split horizontally	:sp	Ctrl+x 2
Split vertically	:vsp	Ctrl+x 3
Switching windows	Ctrl+w Ctrl+w	Ctrl+x o

Source Insight has a standard layout to display project navigation, source code, symbol definition, and function's references in specific windows which can be made visible or hidden by clicking designated menu buttons.

File and Directory view

Code browser should have the ability to handle both files and directories within one single view. The idea is to have minimum distraction while you are busy browsing code. It really kills if you have to always switch your current view and go back to shell to find out what all other files are in the same directory. Having an integrated framework where you can view files and browse directory without losing your focus really improves productivity. After all, *nix operating systems treat file and directory as same.

In the integrated file and directory view, there is no visible difference in the browsing window layout and most of the commands work seamlessly with a few special commands handling the directory browsing operations. For example in vim and Emacs, when you press the *<RETURN>* key over any filename in the directory view, it loads the same filename in the current window.

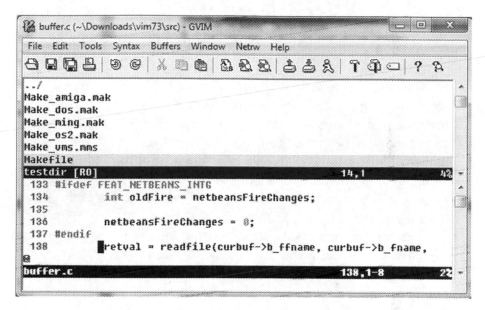

Above screen shot depicts multiple windows opened in the Vim editor with the upper window listing down directory and the lower window displaying the content of a regular C file.

Browsing History

You are looking at some code piece written in *file1.c* and you felt the need to move to a function defined in *file2.c*. All is well if you want to continue working in *file2.c*, but what if you wish to go back to *file1.c*. Still no issues, you can reopen the file in your code browser, but what if you didn't know the name of the last file? Wouldn't it be nice to have the code browser maintaining some form of browsing history similar to web browsers using *Next* and *Prev* buttons! Maintaining browsing history really becomes a must-have feature where we have a huge number of files and it is hard to remember each and every file you last opened.

Vim doesn't provide an exact replica of history maintenance like the one you are used to, but there are ways to achieve the same. Use *Ctrl+o* to move to the previous location and *Ctrl+I* to move to the next location. Events like search, substitutes and marks are constantly saved in a jump list which contains the 100 latest such jumps. Use command ":*jumps*" to list down all such locations. You can also use the ":*ls*" command to list down all the open files in the current window and then use ":*b3*" to jump directly to the third file in that list and so on.

Emacs also has a marker ring similar to vim, within that ring it stores as many as 16 previous marks and you can move to your previous mark by using "*Ctrl+u Ctrl+<Space>*". Additionally, use the command "*Ctrl+x Ctrl+b*" to list down all the open file buffers and press <RETURN> key on the file of interest. A buffer is nothing but a placeholder for an opened file in the Emacs editor. The buffer remains alive from the time the file was opened for the first time till the time it is finally closed. Also, you can use "*Ctrl+x b*" command and provide the number of the file buffer you are interested in. The simplest way would be to use "*Ctrl+x <left-arrow-key>*" and "*Ctrl+x <right-arrow-key>*" to move to the Next and the Previous opened file buffers.

Integrated pattern search in files and directories

Search is yet another important operation while you are having a first look at some huge code base. Search token could be a simple function name, a variable name or a complex regular expression. Whatever code browser you have chosen, it should have the ability to perform search efficiently. Size of files is a big concern while performing efficient search. Another important aspect of integrated search functionality of a good code browser is the ability to perform search over a list of files or directories.

Just press '/' and type your search pattern and press <RETURN> key if you are using Vim. Now iterate through the matched occurrences in forward direction by pressing '*n*' or in the reverse direction by pressing '*N*'. To search backwards you can also use '?' and then type your search pattern. Then use '*n*' to iterate through matches in the reverse direction and '*N*' to move in forward direction. To search the current word under cursor, simply press '*' on your keyboard for forward search and '#' for backward search. Vim also stores search history. After pressing '/' and '?' use the Up and Down arrow keys to access search history.

To search within directories vim provides commands like "*:vimgrep*" and "*:grep*". While "*:grep*" relies on external grep application and its syntax, "*:vimgrep*" is a part of vim itself.

In Emacs press "*Ctrl+s*" and provide the search string. Use *Ctrl+s* to iterate forward and *Ctrl+r* to iterate backwards. To search within multiple files in a directory you can type "*Alt+x grep-find*" and then provide the search pattern.

Hyperlink Navigation

Imagine you are having a look at some function and suddenly you come across a function call. It is quite evident that to understand the algorithm of the current routine, you need to have a look at the function being called. How to search that routine? You can simply use any of the above listed features of a good browser and perform a pattern search. That might take some effort but it will do the job. However, there are better ways to do it. Go to any webpage, and you will see the page is full of links which we call hyperlinks. As soon as you click it, you jump to that new webpage. After you finish reading that new webpage, you can press Prev button to go back to your previous page. ctags and etags can be easily integrated into browsers like Vim and Emacs and they give you the same benefit of hyperlink navigation. You can jump to definition of a function from its function call and later on come back to the same line from where it was called from. It is this mesmerizing feature you will use most of the times in your professional world.

To get to know about the function call references in Source Insight, simply move your cursor over a function name and then do a right click with your mouse. The menu will have the option to jump to the function definition or refer to the calls of the function. You can also jump directly to the definition of a function, a type, a variable by simply doing a Ctrl+= over that text in the source insight window.

Source Insight maintains its own internal database to perform all the operations. However, to achieve the same with Vim and Emacs you need external help in the form of tools like Cscope, ctags and etags.

- **Integration of Cscope with Vim**

  Although Cscope is a great tool to have, but the interface is a bit messy. To go back to the search results if you are already having a look at one of the result, you need to close vim again and again. But, once Cscope is integrated with vim you can utilize all of its functionality directly from the browser.

  You need to have cscope_maps.vim in your $HOME/.vim/plugin directory. It can be downloaded from: http://Cscope.sourceforge.net/Cscope_maps.vim.

  The detailed instructions for integration can be found in Appendix C. Once integration is done, press "*Ctrl+\ s*" when your cursor is over the symbol of interest. You will see Cscope menu with all the options which can be selected by hitting Enter. "*Ctrl+spacebar s*" splits the window horizontally and the Cscope results will be shown in the new window. This is not all, you can also use Cscope in its primitive for by executing commands like ":cscope find symbol *<symbol>*" (or, *:cs f s <symbol>*) directly in the Vim. Check out cscope_maps.vim file for all the shortcuts available or define a new combination of your choice.

- **Integration of ctags with Vim**

  Creation of "tags" file for ctags has already been discussed earlier. If you haven't gone through that then please consider reviewing that section for sure.

  If browser window is already open, you can use command "*set tags=<filename>*" to provide the path to the tags file containing index information. Now, you can press "*Ctrl+]*" to find the definition of the symbol under cursor. Once pressed, the file containing the needed definition will open automatically with cursor positioned at the appropriate line number. To go back to the original position you came from press "*Ctrl+t*". You can always use "*:tag <symbol>*" command to search directly for the definition of symbol *<symbol>*.

- **Integration of etags with Emacs**
  Etags can be tightly integrated with Emacs similar to how ctags can be used with Vim. Use "*Alt+.*" to jump to a tag definition and "*Alt+**" to go back to the original position. If you press "*Alt+.*" for the first time, Emacs will ask you for the location of the TAGS file. Also, you can use the same "*Alt+.*" command to provide your choice of symbol instead of moving the cursor to the appropriate position.

Execute commands

As if we don't have all the facilities of doing almost anything while browsing code, we still have scope for more features. Remember the earlier discussion regarding distractions if you were to switch your code browsing view to go back to shell to run some command. So, it is required that the browsing tool itself can run shell commands directly from the window you are using to read source code. It is not very often that you will execute commands while reading code but the feature is a must from the perspective of completeness.

Running shell commands from Vim is quite easy. To run a single Unix command you can execute "*:! <unix_command>*" directly from the browser. To open a new shell within vim use "*:sh*" command on the vim command line. Once a new shell is invoked use *Ctrl+D* to go back to the browser. There is an interesting command in vim that lets you read text directly from a file or command and appends the result after the current cursor position. "*:r textfile*" will append the text from textfile, while "*:r ! ls*" will append the result of "*ls*" command into the current cursor position.

Emacs exhibits the same behavior through command "*Alt+!*", and the result is displayed in a new window. To run a shell directly from Emacs execute the command "*Alt+x shell*".

You can always refer to the reference manual of Vim and Emacs to have a detailed understanding of plethora of options that these monster editor cum browser cum shell interpreter supports.

With this let us take a rest and wind up this discussion on Code browsers. Hope you had fun and see you in next chapter with a detailed and generous description of the code compilation procedure.

Chapter Review

- Browsing through the amount of code in professional environment needs specialized tools and techniques. In simple words, you cannot use your good old tools like notepad, WordPad or manual bookkeeping to achieve the same.
- There are several aspects that can be associated with a good code browser and you should look forward to choose your browser based on these parameters.
- Cscope is one gem of a tool to perform efficient and quick search within a code base. Cscope maintains its own internal database and a cool interface with the added benefit of being a highly used and stable tool.
- Tag file acts as an index to the programming symbols and hence provides one more option to quickly search the programming specific symbols. Like Cscope, it also maintains its own separate database.
- Tools like Cscope and tags needs constant update of their database whenever there is a formidable code change.
- Browsers like Vim, Emacs and Source Insight provide a great interface for navigation not only within a file, but across directories. The added benefit they provide is their easy to remember keyboard shortcuts so that you don't have to bang your head against the huge reference manual of the browser you are using.
- Multiple file views in a single window along with the integrated directory view in the same browser saves us a lot of time. Always having to lose focus for small requirements while you want to focus only on code browsing is a practice which you never want to practice.
- A good browser also maintains the history and the list of opened files in some internal buffers for your future references.
- Tools like tags and Cscope work not only in standalone mode, they are best utilized when integrated with good browsers. Once integrated they provide amazing hyperlink navigation.

Code compilation & Output viewing

"Practice makes a man perfect" - The more you write and browse through the code, the more you'll feel comfortable at it. But how do you ensure that your code is correct i.e. error free, be it syntactical, logical or run-time? The only way to determine that is to compile your code and verify the output against various test-cases.

You might have been using Integrated Development Environments(IDE) like Turbo C++, Dev C++, Visual Studio, or Eclipse, where all you had to do was to click on "compile" followed by "run" or use the respective shortcuts. Occasionally you may have used command line to compile the code - executing a simple "cc" or "gcc" command and generating an a.out file for execution. As you would have compiled either only stand-alone files or a few of those together in a folder without myriad dependencies, you never had to care about various compiler options and other elements. Generation and application of Makefiles, static and shared libraries might currently sound alien unless yours is a rare and exceptional case.

You already know the program compilation steps i.e. Preprocessing -> Compiling -> Assembling -> Linking; we are not going to discuss about those. This chapter intends to make you familiar with the various elements of Compilation and Linking, for example Makefiles, Libraries,

Dependencies, Errors and Warnings. You shall also learn about viewing output in case of processes and daemons.

All set? Here we go.

Compilation

Code compilation is basically the process of translating source code into object code. Let's run through the various elements that are involved in code compilation-

- Source code files
- Header files
- Object code files
- Libraries
- Makefiles

Source code files

In simplest terms, it is the collection of human-readable set of computer instructions written in a programming language. It is a before-compiled version of the computer program written by someone or generated using the tools for the purpose. The source code files have specific extensions that denote the programming language used e.g. .c or .java

Header files

The seemingly not-so-important header files that just contain the declarations and macro definitions have the information required by several different files or functions. These files allow programmers to separate certain elements of a program's source code into reusable files. The usage of a header file is requested in the program is by *including* it. In C, it is done using pre-processor directive **#include**. Header files serve the following two purposes:

- **System header files** declare the interfaces to the parts of the operating system. Including these header files in your program supplies the definitions and declarations to system calls and libraries.
- **User generated header files** contain the declarations for the interfaces between source code files of the program. Definitions

or declarations that shall be required across the different source files should be updated in the header files. It makes updating them faster and less-error prone. Whatever uses that header file shall automatically use the new version upon recompilation.

Object code files

Object code (.o or .obj) files are the after-compiled version of the computer program. It contains the sequence of instructions that the processor can understand but is difficult for a human to read or modify. An **object file** contains a relocatable format machine code that is usually not directly executable. Object files are produced by an assembler, compiler, or other language translator, and used as input to the linker.

Note: For scripting languages i.e. non-compiled or interpreted programming languages, such as Javascript or Shell script, no object code is generated and there is only one form of the code which is directly executed.

Libraries

A library is a file that contains or rather groups several object files that can be used as a single entity in the linking phase of a program. The code portion that is considered refined and stable over a period of time and does not undergo frequent changes can be built into a library. More about libraries and their types is explained after linking as the description shall make more sense then.

Makefile

Compiling source code can be tedious and the perplexity of the process rises as the size of the code, dependencies among files involved and the span across subdirectories increases. Makefiles provide a simpler and better way to organize the code compilation and are must for a big project. The following content barely scratches the surface of what is possible using make, however it aptly serves as a starters' guide to familiarize with the makefiles so that you can easily understand them and can even start writing your own.

Makefile - a detailed analysis

First of all, makefiles are used so that:

- You do not have to remember/type/copy long strenuous command(s) for compiling the source code.
- You save tremendous amount of time by compiling only the updated files in a project.

Basic structure of Makefile is as follows:

```
1.  target: constituents or dependencies
2.  [tab] construction rules i.e. system_command
```

If the *target* is out of date with respect to *constituents* or *dependencies*, *construction rules i.e. system_command* are processed. In other words, only if any of the dependencies undergo code change, the corresponding target will be built using the construction rules.

Let's understand this with a simple example consisting of just three files: mainFile.c, funFile.c and headerFile.h in a single directory:

```
1.  // headerFile
2.  void fun(void);
3.
4.  // funFile.c
5.  #include <stdio.h>
6.  #include "headerFile.h"
7.
8.  void fun() {
9.      printf("I love to make it\n");
10. }
11.
12. // mainFile.c
13. #include "headerFile.h"
14.
15. int main() {
16.     // Call the function in other file
17.     fun();
18.     return 0;
19. }
```

Conventionally, you would compile this collection of code using the following command:

```
1. gcc -o makeit mainFile.c funFile.c -I.
```

-I. makes gcc to look in the current directory(.) for the include file. The same can be achieved using this simplest makefile for the task:

```
1. # Makefile 1
2. makeit: mainFile.c funFile.c
3. gcc -o makeit mainFile.c funFile.c -I.
```

Here it is specified that the binary *makeit* needs to be generated using mainFile.c and funFile.c by executing the command mentioned in the next line. Remember that TAB is must at the beginning of a command in makefile.

With this, just executing *make* in the directory shall do the job. However, this makefile provides no improvement over the conventional method except for simply typing *make* instead of lengthy commands.

Now consider this one:

```
1. #Makefile 2
2. CC=gcc
3. LD=gcc
4. CFLAGS=-I.
5. DEPS=headerFile.h
6. OBJS=mainfile.o funfile.o
7. makeit: $(OBJS) $(DEPS)
8. $(LD) -o makeit $(OBJS) $(CFLAGS)
9. mainfile.o: mainfile.c $(DEPS)
10. $(CC) -c mainfile.c
11. funfile.o: funfile.c $(DEPS)
12. $(CC) -c funfile.c
```

This is where using a makefile starts paying off. Only the files changed will be compiled followed by re-linking. If no changes are there, make shall tell that and would not waste cycles compiling and linking the code. Notice the macros CC, LD, CFLAGS, DEPS and OBJS used here. CC defines the compiler to be used, LD mentions linker which is also gcc, CFLAGS is the list of the flags to be passed to the compilation commands, DEPS is the set of .h files on which .c files depend and OBJS refers to the .o files. $(MACRO) specifies the value assigned to the MACRO. See make manpage to learn more about the macros.

The above makefile does save some time but practically it is a redundant makefile because this is bound to grow invariably huge with thousands of source code and header files. Worry not as Makefile provides the option of using a single rule to eliminate redundancy as follows:

```
1.  # Makefile 3
2.
3.  CC=gcc
4.  LD=gcc
5.  CFLAGS=-I.
6.  DEPS=headerFile.h
7.  OBJS=mainfile.o funfile.o
8.  MAKEIT=makeit
9.  $(MAKEIT): $(OBJS) $(DEPS)
10. $(LD) -o $(MAKEIT) $(OBJS) $(CFLAGS)
11. %.o: %.c $(DEPS)
12. $(CC) -c -o $@ $< $(CFLAGS)
```

This one rule replaces all the redundant lines from Makefile 2. '%' is a wildcard that matches a part of filename. Therefore, it states that each .o file depends on the corresponding .c file and the header files defined in DEPS. '$<' refers to the dependency list matched by the rule (in this case - the full name of source file). '$@' and '$^' respectively refer to the left and right side of colon(:) i.e. target and dependencies (mainFile.o and mainFile.c respectively).

In this makefile, you may also add the rule for clean by appending the following to the Makefile 3.

```
1.  clean:
2.  rm -rf $(MAKEIT) $(OBJS)
```

Upon executing *make clean*, make shall clean up everything.

By now you should be able to understand and write makefile(s) for small and medium sized projects. However, in professional world the size of projects is fairly large. Such large projects span across multiple subdirectories and thus your makefile should change accordingly. One such code base may have structure like this:

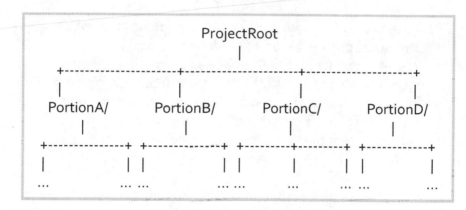

In such cases, a widely used classic approach is to place a makefile at each subdirectory level which compiles all the code in it including the subdirectories. At the top or root, the makefile may look like this:

```
1.  all:
2.  cd PortionA; make
3.  cd PortionB; make
4.  cd PortionC; make
5.  ...
```

As each line in a make target runs in its own shell, therefore there is no need to go back to the previous directory using "*cd.*"

🔊 Makefile is an integral part of any professional code snippet. We only covered the basics to let you get started, the subject is pretty complicated and requires a thorough understanding from its manual

Linking

Compilation does only half of the job. Linking takes the baton where compilation leaves. It refers to the process of creating an executable file from multiple object files.

In case you wonder why there are two different processes: compilation and linking, it is due to following reasons:

- First, probably because it is easier to do things that way because the whole process becomes modular.
- A great advantage is that it allows for creation of huge projects without having to redo the full compilation everytime a file is changed. Just the changed file is re-compiled to generate object files and llinking is directly performed.
- Another advantage is that it also makes it simpler to implement libraries of the precompiled code.

Unlike compilation, the output of linking is something that user can run. During compilation, if the compiler is unable to find the definition for a particular function, it simply assumes the function to be present somewhere and proceeds. On the other hand, linker tries to find the references for the functions and if not found reports about undefined functions. Linking is of two types, static linking and dynamic linking.

Static linking

This is the result of the linker copying the actual code of all the library routines used into the code section of the executable. It has many benefits, the most significant one being assured that all the libraries are present and they are correct version. It is faster and more portable as it does not require the presence of any extra library where it is run. However, for the same reason it inflates the size of the executable. For any change in code or update in library, whole thing needs to be recompiled and replaced.

Dynamic linking

It is accomplished by placing the name of shared library in the executable. Actual linking happens when the executable is run and both the executable and the library are placed in memory. Unlike the static linking, here libraries are not the part of executable and hence the size of the executable is smaller. This saves the disk space and launch time.

With dynamic linking, It is the responsibility of the user to make sure that libraries and their correct version is present in the machine where executable is run. Biggest advantage of dynamic linking is that multiple programs can share a single copy of the library. Also, as the library and the executable are not packaged together so either can individually be updated and replaced. It may affect portability as multiple files need to be copied.

Note: With Static linking, it is sufficient to include only those portions that are directly or indirectly referenced by the executable; however in dynamic linking, entire library needs to be loaded as it is not known in advance which all functions shall be invoked by the executable. The significance of this advantage depends upon the structure and size of the library.

More about Libraries

Libraries are used in linking. As mentioned above, libraries are just the collection of various object files. The benefits of using libraries are:

- They allow program(s) to be more modular.
- They reduce the code size and hence compilation and recompilation time.
- They simplify the multiple use and sharing of the software components among applications.
- Each and every object file need not be stated while linking, instead the individual library can be referenced.

Functions which are to be shared by more than one application should be compiled and bundled into a library. The code put into a library should be stable and robust so that it doesn't negatively affect the applications that use it. Corresponding to static and dynamic linking, there are static and shared(or dynamic) libraries

Static libraries

As evident from the name, they allow for the static linking in a program. A static library becomes part of the executable as it allows the user to directly link to the program without having to recompile its code. However, given today's faster compilers and systems, this advantage is not so significant as it once was. They are, however, often useful to the programmers who wish to permit others to link to their libraries but do not want to provide the library source code.

Shared (or Dynamic) libraries

These libraries are used for dynamic linking and are loaded by the programs when it starts. The file is structured so that the code is position independent and can be loaded anywhere in the memory. An application can load and unload the shared libraries as and when required. This is generally done by an interpreter (usually ld.so) that links the required shared object to the application at runtime. As they are not tightly coupled with the applications, they allow to

- update the libraries to new versions while still supporting the programs that use older and non-compatible versions of the libraries.
- override the specific libraries or functions in a library when executing the program.
- "hot-swap" as no need to stop the program running existing libraries to do all this.

Libraries and Linking

Normally a library is indexed and therefore it is easier to find symbols (functions, variables etc.) in it. Hence, linking a program with libraries is faster as compared to linking with separate object files on the disk. Also, there are fewer files to be opened and processed, which further speeds up the linking process.

Building and using libraries

Now that you know about the libraries, you might probably be wondering how to apply what you have learnt. Time for action! With a simple example, you will not only learn how to use static and shared libraries but also how to create your own libraries.

Let's consider a simple example of creating a library that contains most basic code which returns the sum of two integers and using it. (The code is for demonstration purpose and doesn't handle very large integers values properly):

```
1.  int sum(int a, int b) {
2.      return (a+b);
3.  }
```

This code goes into the library. It exhibits one single function that takes two ints, calculates their sum value and returns it. Let us name the file as *codeForLib.c*

The Header File

Let's create a header file (*codeForLib.h*) for our example with following contents:

```
1.  int sum(int a, int b);
```

The Program using the library

This is a sample program (*codeToRun.c*) that shall use the function from library to obtain sum of two integers:

```
1.  #include<stdio.h>
2.  #include<codeForLib.h>
3.  int main() {
4.      int i = 2;
5.      int j = 6;
6.      printf("\nValues : %d and %d. Sum : %d\n", i, j,
7.              sum(i, j));
8.  }
```

Creating a static library

As you know, a static library is basically a set of **object files** that are copied into a single file. The static file is created with the archiver (ar) utility. First, we create object file *codeForLib.o* from the c file *codeForLib.c* as:

```
1.  $ gcc -c codeForLib.c -o codeForLib.o
```

Now, the archiver (ar) is invoked to produce a static library (named **libSum.a**) out of the object file **codeForLib.o**

```
1.  $ ar rcs libSum.a codeForLib.o
```

Please note that the library name **must** start with the three letters lib and end with the suffix **.a**.

Linking against static library
Now the **codeToRun.c** can be associated with the static library as:

```
1.  $ gcc -static codeToRun.c -L. -lSum -o
    staticLinkedExec
```

Please note that the first three letters (**lib**), as well as the suffix (**.a**) must not be specified. The rest of the library name should be prefixed by -l. So, in the above example, for libSum.a, we specify – lSum. "-L." specifies the current location of the .a file.

Running the executable
We now have executable created as **static_linked** in previous step. It can be simply run as "**./staticLinkedExec**" to get the following output:

```
1.  Values: 2 and 6. Sum : 8
```

That's it. Now you know how to create and use static libraries. Let's move on to shared ones considering the same simple sample code.

Creating a shared library
As mentioned above, shared libraries always have position independent code. -fPIC option tells gcc to create object file(**codeForLib.o**) with position independent code as:

```
1.  $ gcc -c -fPIC codeForLib.c -o codeForLib.o
```

Next, the shared library (libSum.so) is generated as:

```
1.  $ gcc -shared -Wl -o libSum.so codeForLib.o
```

Please note that the library name must start with the three letter **lib** and end with the suffix **.so**

Linking against shared library (Dynamic linking)

Now the **codeToRun.c** can be associated with the shared library as:

```
1.  $ gcc -o dynamicallyLinkedExec codeToRun.o -L. -lSum
```

Please note that the first three letters (lib), as well as the suffix (.so) must not be specified. The rest of the library name should be prefixed by -l. For example, in this case, we specified –lSum for referencing libSum.so.

Running the executable

We now have executable created as **dynamicallyLinkedExec** in previous step. The procedure to run the executable as mentioned in many books is to specify the LD_LIBRARY_PATH as

```
1.  $export LD_LIBRARY_PATH=./dynamicallyLinkedExec
```

Now running the executable like we did earlier in case of static linking shall produce same results.

Note: Don't panic in case it doesn't work for you, and throws the following error:

```
1.  /libexec/ld-elf.so.1: Shared object "libSum.so" not
    found. Required by "dynamicallyLinkedExec"
```

It simply means that the interpreter could not locate the library. Create a soft link from /usr/lib to libsum.so as

```
1.  $ ln -s <initial_path>/libSum.so /usr/lib/libSum.so
```

Now, running the executable should work like a charm.

◀⅃) You may successfully be able to compile your source, but never be sure till your linker gives a green signal. Linking errors, though hard to find, needs proper attention.

Compilation alerts

During the compilation process, if everything doesn't go well then the compiler /linker catches it and provides meaningful alerts which can be broadly categorized as:

- Compiler Errors
- Compiler Warnings
- Linker Errors

Compiler Error

An error which is encountered during compilation of the program and prohibits the generation of object code is usually fatal and is classified as a compiler error. Compiler errors are syntactic in nature. Common compiler errors are:

- **Variable Undeclared**
  Either you have forgotten to declare a variable before using it or else you have mistyped as variable name. Remember, C is case-sensitive.
- **No such file or directory**
  You asked compiler to compile something that doesn't exist.
- **File format not recognized**
  The compiler expects the files specified for compiling to have proper extension e.g. .c in case of C code files. Putting C code in code.txt and trying to compile shall lead to this error

- *parse error before 'string'*
  A typo in code, or an unrecognized symbol at the beginning of statement, perhaps some missing parentheses or semi-colons.
- *Character constant too long*
  You have enclosed a string in single quotes instead of double ones. Compiler expects a character in single quotes and string double quotes.
- *Un-terminated string or character constant*
  You do not have balanced single or double quotes.

Compiler Warnings

If compiler identifies that it can produce the object code but there might be runtime issues or the program may not behave as expected, a warning is issued. The warnings are not severe enough to keep your program from compiling. You would have comfortably ignored the warnings encountered while writing programs in college. However, you do not want to ignore them when you work in a corporate. You can tell your compiler to treat warnings as errors by specifying -Werror flag. In professional environment, usually the warning level of compiler is set to highest level to avoid any of the compiler warnings actually turning into bugs during run time. For example, the compiler may warn you about variables not initialized. They may contain garbage value and cause issues. Some most commonly seen warnings are:

- *makes pointer from integer without a cast*
  You passed an integer where a pointer was expected.
- *Assignment in conditional*
  You used = instead of == in an if condition
- *Code not reachable*
  Some portion of your code can not be reached in any case.
- *Uninitialized variables*
  As mentioned, you specified the variable but forgot to initialize.
- *Unknown escape sequence*
  There is a sequence preceded by backslash '\' present in the function which is not recognized. For example \n is a valid escape sequence but \z is not.

Linker errors

During compilation, if a referenced symbol is not found by compiler, it just assumes that it shall be present in some file. Linker does the job of traversing through corresponding object code and 'link' the references. Linker errors are about missing or multiple definitions for functions, structures, global variables. Here are the mostly encountered ones:

- **Undefined symbol**
  You might have declared but forgot to define it. For example, something is encountered that seems like a function call but no such function is found. Other reason might be you wrongly or did not specify the correct path to the corresponding library.
- **Multiple definitions**
  As the name suggests, you defined the same thing more than once in same file or in different files but declared as extern. Other case, you provided multiple copies of the same object file to the linker.
- **Undefined reference to main**
  Main function is not found in the code that you have supplied to the linker. Either you forgot writing main or supplying the file that contains it or are trying to compile code that is not meant to be a standalone executable, for example, a library.
- **Unable to find <some library>**
  The library to be referenced is not in the paths specified.

This list is not comprehensive and only some of the most common errors are mentioned to give you an idea about the same.

🔊 Errors generated by the compiler are there only to prevent you from mistakes done unknowingly. You should try to eliminate them at the coding stage itself. After all, a compiler is a tool to generate object code and not to catch your coding errors.

Compilation flags

Compilation flags or the compiler options are the keywords that can be specified in compilation command to control certain aspects. These aspects could be the efficient use of the compiler, warnings and errors reporting, nature of the load module generated, or type of printed outputs to be produced, optimization of the generated binary and so on. Any industry standard compiler like gcc provides thousands of options that you may find in the man page of the compiler. Following is a list of few important flags or options that are extensively used in the industry and would be quite sufficient in most of the scenarios:

Output options

- -c : Compile the source files but suppress the linking part. The compiler output would be corresponding object files for each source file.
- -o <executable_name> : Generate the executable with name specified instead of a.out.
- -D<macro> : Define a macro. Can use -Dmacro=value to assign the value to the macro. The macros might be used inside source files for some code that should be executed only if the macro is specified.
- -pipe : Use pipes rather than temporary files for communication between the various stages of compilation.

Language options

- -ansi : Enforce the ANSI C standards.
- -traditional : Support some aspects of traditional C compilers.
- -fno-strict-prototype : Functions may be there with no arguments.

Warning options

- -pedantic : Issue all warnings demanded by strict ANSI standard C or specified standard.
- -w : Inhibit all the warning messages.
- -Wimplicit : Warn if type is not specified in declaration or a function is used before being declared.
- -Wunused: Warn if a local variable is defined but not used.
- -Wall : Show all the reasonable warnings.
- -Werror : consider all the warnings as errors.
- -Wmissing-declarations : Warn if a global function is defined without a previous declaration.

- -Winline : Warn if a function specified as inline cannot be inlined.
- -Wunreachable-code : Warn if some part of code can never be executed.
- -Wswitch-default : Warn if a switch statement doesn't contain a default case.
- -Wfloat-equal : Warn if floats are being compared for equality.

Debugging options

- -g : Produce debugging information. Causes the executable size to increase.
- -ggdb : Produce debugging information in native format.
- -pg : Generate extra code suitable for gprof analysis.

Optimization options

- -O or -O1 : Optimize compilation. It takes somewhat more time and lot more memory for a large function. Reduces the executable size and execution time.
- -O2 : Optimize even more. Includes O1 plus almost all optimizations that do not involve space-speed tradeoff.
- -O3 : Optimize yet more. Includes O2 and turns on -finline-functions.
- -Oo : Do not optimize.
  Note: *If you specify multiple '-O' options, last option specified will only be effective.*
- -finline-functions: Integrate simple functions to their callers.

Directory options

- -I<dir> : Append directory 'dir' to the list of directories to be searched for include files.
- -L<dir> : Add directory to the list of directories to be searched for libraries.

Linker options

- -l<library> : Link to a standard library.
- -nostdlib : Do not use the standard system libraries and startup files for linking. Use only the ones specified by user.

- -static : Prevent linking with shared libraries on systems that support dynamic lilnking.
- -u <symbol> : Pretent that the symbol specified is undefined. This forces the linking of library modules to define it.

Output viewing

In case of processes or rather say foreground processes, it is what you are already familiar with. The process shall execute and dump the output on the stdout or stderr unless specified to a particular file in the program. If the output is too huge, you may redirect it to some file as:

```
1.   ./someprocess > /tmp/somefile.txt
```

This file can be opened and viewed it in a text editor of your choice.

Daemons

Daemons are the programs that run in the background, waiting for the events to occur and offer services. They are not under the direct control of an interactive user. Typically daemon names end with the letter d, for example sshd, ntpd. Generally they keep running unless killed by user or system is shut down. So, how to see their output?

Well, in case of daemons the output is either seen on screen, in logs or it is piped out which can be redirected to a particular file and observed using tail -f as the file keeps growing. The logs are usually collected under "/var/log/" directory. As the daemons keep running continuously, most of the logs are enabled only when you need to debug. Some daemons require you to restart them with --debug option or some similar predefined flag; other daemons may provide for a way to enable the debug logs without restarting by sending a signal to the daemon which internally sets the corresponding flag in signal handler.

The logs are even categorized into various verbosity levels; greater the level - more detailed is the information collected. This is very useful in debugging where you may want only particular kind of logs to identify the issue and avoid going through thousands of lines which would not be necessary for the situation. For example, sshd provides for QUIET, FATAL, ERROR, INFO, VERBOSE, DEBUG or

DEBUG1, DEBUG2 and DEBUG3 as various verbosity levels where DEBUG3 provides for highest level of verbose output. You may only want to see what error is being thrown, you can get it by enabling only the ERROR logs.

So, if you want to debug the sshd daemon, you need to follow the following procedure:

- kill the sshd daemon: Get the sshd pid and kill the process using following command set:

```
1.  $ ps ax | grep sshd
2.  $ kill -9 <pid>
```

- Restart sshd with debug flags as:

```
1.  /usr/sbin/sshd -dddf /etc/sshd_config
```

This shall enable the level 3 ssh debug logs. -f option specifies the config file to be used.

- Now open a new terminal where you connect to sshd. The terminal where you restarted sshd with debug flag shall dump the logs on screen which will look something like:

```
1.  debug2: load_server_config: filename
    /etc/ssh/sshd_config
2.  debug2: load_server_config: done config len = 163
3.  debug2: parse_server_config: config
    /etc/ssh/sshd_config len 163
4.  debug3: /etc/ssh/sshd_config:21 setting Protocol 2
5.  debug3: /etc/ssh/sshd_config:110 setting Subsystem sftp
    /usr/sbin/sftp-server
6.  debug1: sshd version OpenSSH_4.7p1
```

Note: *If you want the debug output to be logged in a file, you may specify the same in sshd_conf file. Specifying "FacistLogging yes" in the conf file prints the debug messages in the system log file(typically /var/log/syslog in *nix systems).*

This section should encourage you to develop an important aspect of writing code - providing enough comments and log messages for run time analysis. Imagine being trapped in a situation where a daemon has stopped responding and there is no way to determine its running state using run time logs. Let us conclude this chapter with this noble thought of providing enough information to the software community that will be using and maintaining your code in future. Let's meet in the next chapter that will describe the logical and physical arrangement of a large code base.

Chapter Review

- Compiling code in a professional environment is altogether a different business. You need to do much more than simply pressing the compile button in the compiler tools such as Turbo C.
- Conversion of source code written accordingly in programming language syntax to the binary format (object code) that your underlying operating system can understand is the primary task of the compiler.
- Creation of object code may require more than one source file. These files depend on each other in a specific way to generate the final object code.
- Makefile is a placeholder to specify the rules and dependencies that facilitates the generation of object code from given set of source files.
- You need not remember all the rules and dependencies while compiling the code every single time if you are using Makefile. Other benefit of this approach is to avoid the typing mistakes which you would make if you were to manually type the commands every time.
- You need not compile all the source files if you haven't changed all of them. Makefile takes care of this by only compiling a file if any of its dependent changed recently.
- Linking is the process of creating the final executable file from the available list of object code files. The executable thus generated can be run by the user and hence is available for testing for various input sequences if it accepts any.
- Linking can be done in two possible ways - static and dynamic. If we have whole of the library routines present in the main executable's code section it is known as Static linking. Whereas, in dynamic linking we only keep the name of the shared library which will be loaded as and when required at the run time.
- Libraries are files generated from a combination of one or more object files. Libraries are known as shared or static libraries driven by the fact that whether the linking was dynamic or static linking respectively.
- Compiler and Linkers also warn us of the conspicuous mistakes done while coding. Depending upon the severity we can get

warnings or error. While warning doesn't stop the compilation process, error doesn't let the compilation process succeed.

- A Daemon is a process running in background and thus you cannot directly interact with them.
- You need to have abundant and meaningful debug messages in your code especially if it is going to be used as a daemon as they might serve the useful purpose of understanding the runtime state of the process.

4

Understanding Code

So now you how to write good code, gliding through it, and compiling plus linking it like a professional. It's time to delve an aspect of coding that troubles almost all fresh professionals and even the seasoned ones – Understanding code.

Imagine thousands of source code files spanning across hundreds of directories, with each file containing several thousand lines of code. What if you are asked to go through this code and try to understand it in a week or less. Sounds freaky?

This is a genuine scenario that otherwise comes as a shock after you join the company. Every product has enormous size of code written and maintained over years. Like it or not, you shall be acceding this 'heritage' and shall obviously have to understand it before working on it. It'll be great if someone may help you but everyone in the corporate world is busy with their work and no one shall have time to spoon-feed you. By the way, it's not the only reason why this chapter finds a place in this book.

Here are few more:

- You sure would have had this 'wonderful' experience of trying to understand the code written by someone else and feeling like,

"Aarrrrrgh!! You evil coder, I must curse you to die a very slow and painful death". When you have to deal with the code written and maintained by hundreds of employees from across the globe over years, the agony and difficulty level can only increase.

- You just don't have to work with the plain code but various integrated libraries and third-party components as well. Some portions of the code might even be written in a different language.
- Furthermore, code spanned and cross-referenced across multiple directories and many levels of subdirectories does bewilder fresh professionals, who are otherwise accustomed to using a single directory to dump all the code files.

Typically a person takes months to get hold of the code because he starts with understanding it in the traditional manner, exactly how it was done in college. This chapter is written to help you to speed up this process by adopting a faster alternate approach which otherwise is learnt only through experience. This chapter shall enable you to whiz through even the gigantic code-bases in far lesser time. It shall also help you in understanding and remembering the flow much better.

Sounds cool? Read on.

Logical layout of Source code

Suppose you go to a library containing a huge collection of maps of almost all the cities in the world and your task is to find out the map of New Delhi. How would you go about it?

If the person maintaining the maps in the library is a bad organizer, God only can save you as you go about looking out for the map of New Delhi amongst the haystack of thousands of maps. But, if the librarian is a good intelligent gentleman and he likes to keep the maps in a logical ordered manner – you can find the required map in minutes. One such logical order could be to maintain the maps in the hierarchical order of continent -> country -> state -> city. Along with this the maps in the individual category can be maintained in the Alphabetical order to facilitate the search procedure.

The same analogy can be applied to the situation when you sit in front of your machine looking at the huge collection of source code files for the first time. At times, the number of files can be much larger than the maps of cities you can find in a library. To worsen the problem – you don't even know which file contains what piece of code. In the case of maps, at least you know that you have to search for map of New Delhi. Chaos everywhere, isn't it?

However, the situation is not that bad as it looks from the surface. Due to stringent coding guidelines and coding conventions followed by any good professional organization, you will rarely find an un-organized layout of the source code at hand. Projects where every new piece of code being added is supervised by peers through code reviews offer a very good chance that the code is maintained in a proper logical order. Not only it produces a cleaner coding environment but also helps the newly joined member to get their hands dirty with code at the earliest. This clearly shows that if the source code files are properly structured at logical boundaries, it becomes easier to understand the code.

There could be many logical groupings which can be applied to the given set of source code files to facilitate understanding its functionality. Let us be proactive and instead of finding out the ways the code is organized and put at our disposal, we can define the rules that we would like to set if we were the original writers of the same piece of code. The best way to understand a written piece of code is to refrain from reading it.

A better way is to put yourself in the shoes of the person who originally wrote that piece of code and then figure out at least one possible solution for the problem. You may not always be able to do that but even if you have given the problem a try, it is worth it. Once you are ready with your little thought over the implementation of the code, then read the actual written code and you will understand the code better as you already are familiar with the problem and the possible approaches to solve the problem.

Picking up an example would be the best way to step forward. Let us discuss how we would like to arrange the source code of a simple chat program. You can write the program in one single huge file "main.c", compile it, and you are done. However, that is not the professional way of doing it. Even for such a small code base, you should always give preference to arranging code in a modular fashion rather than worrying about spending a little extra time in achieving the same. To make the code modular you need to spend time in the design phase itself and carve out the logical partitions that can be molded into separate modules. For a chat program, the logic can be segregated into different modules based on the functionality of the software. On the superficial level, one can always have two separate modules – one each for client side code and other one for server side code. Once this is in place, you can always keep one separate module for utility functions e.g. string manipulation functions can go in a separate file say "*strutils.c*". Similarly the code can be spread across multiple *util* files, where one file contains only a specific piece of utility code.

With basic segregation done, how about writing some generic code to handle connection management in case a chat client wants to communicate to a server or a server wants to dictate few commands to the clients. It calls for writing generic code that can be extended for future needs as a different module. As soon as a new client logs in, you need to authenticate the user using some authentication mechanism – which itself can be molded into separate library (module). Then there are logging mechanisms, various timers which can be independently written into libraries. Thus you can actually write almost all of the independent code sections into different sub-sections which we have been referring to as modules (or libraries). The relationship between these modules can best be described as shown in Fig 4.1.

Now, that you have a well-designed and modular code to maintain –

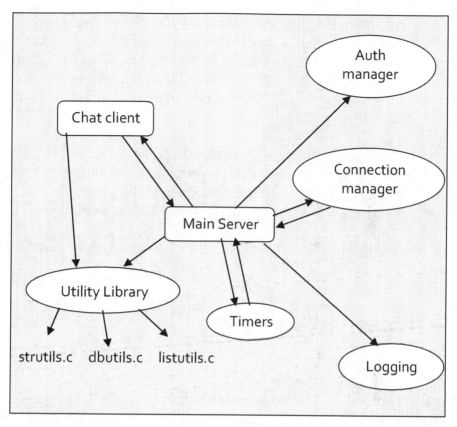

Fig 4.1 inter-module interaction within chat software

How do you get benefitted from this exercise? After all, you are here to understand someone else's code. Perhaps you have already understood the point being made here. If not, here it is-

Considering the importance of proper structuring of code, one can presume that whatever piece of software he is going to work in any professional world will follow the same pattern. In other words, the point being enforced here is that the well-established code bases are actually based upon modular approach that has been in discussion thus far. By taking this aspect of code management at your advantage, you can easily understand the code flow and identify the relevant source files. Instead of blindly fiddling through the plethora of files, you can pinpoint to the location of the files you are interested in, almost similar to the example of finding the map of New Delhi in a properly organized library.

Code organization patterns are not the only means to help you guide through the code. Most of the files and directories follow a certain naming convention which is a very good indication of the code those files / directories contains. Few such vital aspects that you can keep in mind while exploring any source code base for the first time are summarized below:

- Logical layout of the project based on the functionality.
- Physical layout of the project in terms of its directory structure
- Naming convention of the files and directories
- Using code browsing tools to their maximum potential

The rest of the chapter is centered around these points explained in detail.

🔊 To understand already written code, first think of the possible ways you would have written the code if you were the original developer.

Logical layout of the source code

The importance of having a logical grouping for the source code in the form of modules has already been discussed in the previous section. This section extends that concept and applies it to a real world example to explore possible building blocks of any large software program.

Chat software written in C would easily pass as an appropriate example here as you have already gone through it in previous section. The immediate question that needs attention is – where to begin? You will be having tens of files containing the source code in front of you, but out of them which one to pick first and get going. In similar situations, following a top-down approach has always worked for most of the experienced developers. In the context of software programming "Top-Down approach" is just a fancy name given to the process of writing the top-most wrapper routines first and then move down to writing low level specific routines. The same approach can be applied in understanding a large code base.

The chatting software can be broken down into two logical pieces — server side and client side. Server is a daemon which will always be running while multiple clients can be run as and when required. This means, both the client and the server should be having a separate *main()* function so that they can run as a separate entity. Now, it's time to step inside *main()* routine and see where all you can go following a "top-down" approach from here.

Main components of the main() routine present in the chat server can be listed down as:

init_server()

An initializing routine whose primary job is to start all the threads, get the server ready for interaction with the clients. It would be this function's responsibility to initialize the data structures, file descriptors, sockets etc. Setting up the environment for logs and debug info would also appear in the initialization phase of any well-thought chat server.

A big fat while loop

program running at the server side normally runs as a daemon, which means that it will be running forever once started until it crashes or someone explicitly kills it (e.g. sending a signal via *Ctrl+C*). To simulate the behavior of a daemon, a loop is needed. So, the next important component of your chat server program will be the big fat loop. It may run indefinitely, or till a timer expires, a particular signal got caught, no more data on the socket we are listening or any other error condition. Most of the functionality of the server is repeatedly performed within this loop.

Connection manager

Next meaningful component has to be the core server module that handles the client connections. The server might be having a special thread that will be simply listening to some configured port for the incoming client requests. Once such a request is received, server needs to authenticate the connection and send some acknowledgement to the client informing it that a valid connection has been established with the server. Simultaneously the server will allocate the resources for this new connection.

Handling client requests

Once an authenticated client connection is established, the server might start one worker thread to handle the client requests. In this case it will start a dedicated connection channel between the two chat clients that wish to communicate amongst themselves. Also, if clients wish to terminate the session this handler needs to send back the intimation to server which will free the resources and do some book-keeping.

Error handling and logging

An important aspect of any stable and popular software is to have a robust error handling mechanism and plenty of debug logs to help in analysis if anything goes unplanned. The server needs to maintain enough debug logs along with proper error messages to help it and the clients if it is not able to process client's requests. Server also needs to maintain statistics and counters as a part of its debug functionality.

Have a look at Fig. 4.2 on the next page which outlines the bare-bone structure of the modules and sub-modules discussed above. The sample program lists down the major sub-modules in the form of function calls only but it should give you an idea as to how it would really be in a professional environment. Also note the pointer *pserver* which contains the server context throughout the life of the program which is only freed at the very end.

```
1.  int main ()
2.  {
3.      int rc = 0;
4.      exception_t *e = NULL;
5.      server_context *pserver = NULL;
6.
7.      _try {
8.          // initialize the server resources
9.          _call(e, init_server(pserver));
10.
11.         // Ctrl+C etc.
12.         _handlesignals(pserver);
13.
14.         do {
15.             // connection handler
16.             _call(e, handleconnection_serv(pserver));
17.
18.             switch(pserver->client_reqtype) {
19.             case CLIENT_NEW_CONN:
20.                 _call(e,
21.                     handle_new_conn_serv(pserver));
22.                 /* similarly other calls */
23.             }
24.         } while (!rc);
25.     } _catch (e) {
26.         // handle exception and update error codes
27.         updateErrorCodes(pserver);
28.     }
29.
30.     free_server(pserver);
31.     return rc;
32. }
```

Fig 4.2 outline of main() routine at server side

Similarly outline of *main()* routine for the chat-client can be broken down into following sub-functions:

init_client()

Allocation of resources like data structures, sockets, logs etc. takes place here. Integral step in the initialization process is the loading of client specific configuration to do the customization in the look and feel of the client UI e.g. load the client window with the same dimensions which the user would have used in its previous chat session.

Connection handler

Clients need to connect to server using the specifications laid down by the server (e.g. the port number on which the server intends to listen for new connections). It authenticates the login credentials specified by the user.

End up in a loop

After successfully connecting to server, client ends up in big loop where most of the functionality lies. User can now see a list of peers with their availability for live chat with whom he can initiate a chat session. To handle this, within the loop a handler is needed to capture user's requests.

Start chat session

Once request from peer is received, a new worker thread is invoked to handle it. A request to server will be sent with peer information. Server will again do authentication, book-booking and if everything goes well, client will receive information of the established channel. Two peers can chat and later on terminate session using the same channel information (e.g. session token).

Receive chat request

Contrary to starting a new chat session, a handler will be running to receive chat requests from other peers. Once such a request is received from server with some meta-information of the peer, user needs to accept or decline the request. If the user agrees to this, a new thread would be invoked to handle the chat session. Later on the connection can be terminated by sending a request to server.

Log out

All the allocated resources need to be freed once a logout request is received. Some statistics and logs will be collected too, along with a request to the server to close the open connections to this client and all the connections opened with other peers. Finally, user's chat window will be closed which may be augmented with a sign-out message.

```
1.  int main ()
2.  {
3.      int rc = 0;
4.      exception_t *e = NULL;
5.      client_context *pclient = NULL;
6.
7.      _try {
8.          // initialize the client resources
9.          _call(e, init_client(pclient));
10.
11.         // Ctrl+C etc.
12.         _handlesignals(pclient);
13.
14.         // connect to server and authenticate
15.         _call(e, connect_server(pclient));
16.
17.         do {
18.             // update user chat window, show all peers
19.             _call(e, updateUI_cli(pclient));
20.
21.             // wait for user side inputs
22.             _call(e, getUserRequests_cli(pclient));
23.
24.             switch(pclient->reqType) {
25.             case NEW_REQUEST:
26.                 _call(e,
27.                     initiateNewChatReq_cli(pclient));
28.                 break;
29.             case JOIN_REQUEST:
30.                 _call(e,
31.                     joinChatSession_cli(pclient));
32.             }
33.         } while (!rc);
34.     } _catch (e) {
35.         // handle exception and update error codes
36.         updateErrorCodes(pclient);
37.     }
38.     Free_client(pclient);
39.     return rc;
40. }
```

Fig 4.3 outline of main() routine at client side

Physical layout of the source code

Logical layout of the source code helps you understand the different modules or sub-modules that comprise it. It aids in building a thought process that can help you decode the organization of the code on conceptual basis. This section deals with the physical layout of the source code and how it can augment your thought process in understanding the source code better. Although these two sections run in conjunction with each other, however we have chosen to deal with them separately because they solve a different purpose.

You must have heard of "cause and effect" theorem. The logical separation of a piece of source code is the cause that affects a certain physical layout of the same. Thus, the physical layout is purely driven by your logical splitting of the code.

Similarly, you can take the necessary help from the previous section and try to see how to design the directory structure which is quiet self-explanatory. By self-explanatory it is meant that it should be so obvious that even a new comer like you would know what code lies beneath that file or directory.

The very first physical separation of code would be to have different directories for server and client. Along with this you would require separate placeholders for code used by both the modules in "common" directory. Various utility functions should go to their corresponding files and ultimately kept under "utilities" directory. Another useful practice to follow is to have all the header files kept under a separate directory named "include". Instead of having multiple header files scattered here and there along with the source files, it is always good to move all of the header files in a separate folder and keep them at a location where the peer folders can access them with ease.

Modules like error handling and logging are not specific to any module and thus could be kept at the top-most level only. This avoids maintaining duplicate code and also helps future developers and newly joined members to almost understand the module without the help of the original author of the code. Logically these two modules can be kept either in "common" or in the "utilities" folder. Let us choose utilities for now and move on.

Modules like connection management, authorization, handling chat request etc. are specific to either server or client. Thus, they should be arranged appropriately under server or client folder. There would be few cases where you would find that the same piece of code or function could actually suffice for both of them and hence there is no need to maintain them separately. A very common example would be the function that frees / allocates memory to resources. You can very easily have a common routine and pass one flag to indicate whether the call is from server context or from a client context. This will work fine and definitely avoids duplicate code to be written and maintained. But, the counter-argument in this context is one factor that always overrides other factors – "*code readability and modularization*".

Using a single routine might not be scalable in future. Who knows, all of a sudden the management decides to rewrite the whole server part of the code. In that situation you need to search throughout the complete source code for the instances wherever a flag is used to multiplex two functions into a common routine. For someone who is relatively new to organization it is not intuitive to follow such coding practices. Hence, it is always better to go for a code piece which is easily readable rather than polluting it with bad programming practices which go against the code modularization process. A modular code is something which will tell on its own the various sub-modules it contains rather than you aggressively trying to find it out.

Within each directory, the files should be carved out such that one single file should handle the code specific to a single feature only. Even though this might lead to a lot many files having just one or two functions, but it is worth the pain. While writing or designing something new, always remember that apart from you there will be hundreds of other developers who will be working on the same code section and thus it should be readable enough to convey the coding strategy clearly to all of them. And it is not always for the others, even you will face difficulty in understanding non-readable code written by you (after few months for sure).

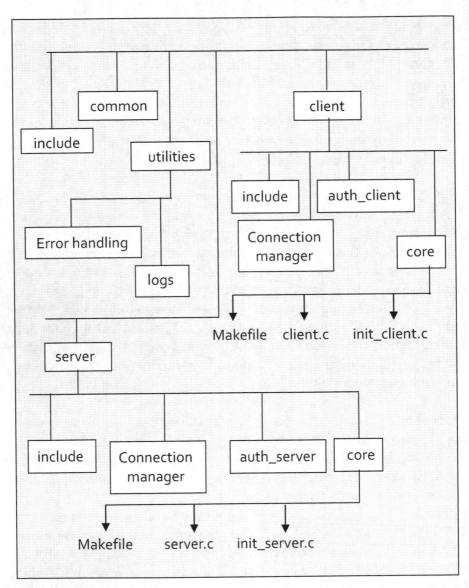

Fig 4.4 Outline of a sample physical layout of client-server chat program

Naming Convention for files and directories

Now that you understand the logical and physical distribution of large code base into smaller modules and sub-modules, half of your job is done. You can easily figure out the location of the routines or files or directory where a particular coding logic is written. This is far better situation than the one with which you started before this chapter. But, it is not done yet. Good developers always leave hints and markers for other users. One such hint is in the form of naming conventions adopted by the developing community while designing the names of the files, directories and functions.

- Directories have the same name as the modules they contain. If not the same name, then definitely some prefix or suffix to help you out. If you go through figure 4.4, "*auth_server*" and "*auth_client*" are named like this to let you know that they are dealing with authorization code for server and client respectively.
- The name of the generated binary file also conveys the name of the module. Along with this normally the binary file name also conveys whether it will be run as a daemon or not. For example, "*serverd*" tells that this is a binary executable of server module and it will run as a daemon (note the suffix −'d').
- Same goes for the file names. Names of the file can indicate the kind of functionality provided by them along with the main module information. For example, "*init_server.c*" and "*init_client.c*" clearly indicates that they contain the initialization routines for the server and client respectively.
- Function name can also provide the name of the main module and its functionality through the use of suffix and prefix.
- Similarly, all the common and utility routines and files would have "common" and "util" prepended or appended in their names.
- Even the data structures would give an indication of the modules or files they are used in. For example – "*server_context_t*" specifies that this structure contains the context information of server at any given time of execution.

Going forward, when you look at the name of a function or a file give due importance to it and try to extract what all extra information it can provide. This step is crucial to nail down your search for the piece of code you are interested in.

Using Tools to understand code

Hopefully now you understand where to look for what you are interested in. Why not use some specialized tools which can get you there quickly.

Cscope

Cscope has its specialty in quick search for a function, file, header files and other C symbols in manageable time. Applying the information gathered in this chapter through Cscope would really augment your understanding of any huge code base.

Tag Search

ctgas (for vi) and etags (for emacs) has been discussed in detail in previous chapters. The beauty of tag search when properly integrated with editors like vi and emacs is the ability to browse the code in a hyperlinked fashion. This greatly saves your time if you want to jump from one function to another to get a better understanding of the code.

Debugger is your best buddy

Debugging a program with a good debugging tool is nothing short of heaven. This is one tool where you can understand the code by executing statements one by one. Debugging is a vast and interesting topic in itself, and therefore the next chapter is dedicated entirely to it.

◀)) Building good understanding of code requires the knowledge of what to look for, where to look and how to fetch that information.

Chapter Review

- It is always better to be proactive and think of the design as to how it should look like, instead of trying to decipher the code by just looking at it.
- Logical layout of code refers to the disintegration of one huge code base into various modules and sub-modules depending on the functionality they provide.
- As an example - a chat software can be broken down into two major components. A server side and a client side. There would be common pieces which can still be arranged as common and utility modules.
- Physical layout goes on parallel with the logical layout. The only difference is it helps in understanding the distribution of code on the disk. It refers to the actual arrangement of source files, header files, binaries and directory.
- To help you further in your path of building understanding of code base, you can follow some naming conventions to unravel the underlying coding logic.
- Once you are aware of the different building blocks and the way a piece of code is arranged, you can use tools like Cscope, tags and debugger to help you further in the cause.

5

Dexterous Debugging

A famous Latin proverb says, "*Wise men learn from the mistakes of others, fools from their own*". Here is the updated version, "*Wise men learn from the mistakes of others, fools from their own, and engineers are the ones who fix them; in the field of software, software engineers to be more precise.*"

Jokes apart, making mistakes (read bugs) are part of being human. As we humans continue to create stuff, mistakes are bound to happen. No single software ever written has been or will be perfect. Usually the Version 01 is not even great. Then engineers like us continuously work hard to find errors, flaws, faults, failures, areas of improvement and conditions where it goes all crazy. Collectively, we term all of these as bugs. The best part is that they can be, and are corrected!

As the fixes go in to rectify the identified bugs, new ones get introduced. *Aarrgh! so much for being human*. A major portion of a software engineer's work life goes in identifying and fixing bugs, or debugging. Worry not, because it is quite interesting and somewhat like being Sherlock Holmes in the digital world trying to discover clues and making sense out of aftermath to identify what caused it.

Writing a software from the scratch or maintaining or enhancing the already existing one, you will almost daily face situations where you have to debug. In other words, in your quest to become distinguished software pro, debugging is one skill you have to master no matter what. Therefore, it's not exaggerating to say that this one chapter is most significant of all. As you read through the next pages, you'll get equipped to take on bugs by learning various methods and types of debugging, efficiently and effectively using the tools developed for the purpose and much more. So, fasten your seatbelts and brace yourself, here we go.

The BUG Story

The term "bug", to describe an error, flaw or mistake in a system that results in an unexpected or incorrect behavior, has been around quite some time. Even before the dawn of computers, the term "bug" was popular among engineers and repairmen, Thomas Edison talked about bugs in electrical circuits in 1870s.

Yet is was computer pioneer Grace Hooper (surprisingly sounds like the name of a bug :P), who is accredited with popularizing the term especially with respect to the computers. In 1946, one of the early electromechanical computers (Mark II) was found to have an issue. The cause was traced to a moth trapped in a relay. The bug was carefully removed and taped to the log book. The log book is on display in the Smithsonian National Museum of American History, with the moth attached.

The operators who found it were familiar with the term and, amused, kept the insect with the notation "First actual case of bug being found." This resulted in the widespread use and acceptance of the term in the computer lexicon.

Types of debugging

Already you would have done some debugging during your college projects and become familiar with breakpoints and watchpoints. If not, just know that they are the backbone of debugging process and shall be explained later in the chapter. For now, let's go through a broad classification of the various ways a program can be debugged:

Print debugging

It is the most primitive method of debugging- just add `printf` or corresponding statements in the language being used. They indicate the flow of execution and can print the values of variables. You may as well use `fprintf` to direct the output from these statements to a file which can be opened at beginning of the program and closed before termination. The issue with this approach is that the programmer has to introduce these statements, compile it again before running the binary and then remove those after debugging. It goes without saying that the customers or end users need not and should not be bombarded with those details.

Post mortem or core-dump debugging

It is debugging of a program that has already crashed or abnormally terminated. Such mishaps produce a core-dump or memory-dump file which is a snapshot of the memory at the time of crash. Using debuggers like gdb, the core files can be analyzed to identify the issue. The crashes that happen at customer sites are most of the times analyzed and resolved using this technique.

Live Debugging

Another way is to recreate the environment or conditions in which the issue was observed and debug the program as it runs to see what exactly is going wrong. To do so, the process to be debugged is either started as a child process of the debugger or the already running process is attached to the debugger. Live debugging can also be utilized to quickly understand the flow of the program.

Remote Debugging

It is another form of live debugging. The only difference being that the code is present on a different machine than where the executable is being run. A debug server is started on the machine where the process to be debugged is being run. From the machine where code is present, debugger connects to the server over a network. Once connected, debugger can control the execution of the process remotely and fetch information about its state.

Although, these methods are the direct ones and provide deep insight into the process execution, sometimes *debugger-less* methods fare better. They sometimes consume lesser time to locate possible location of the culprit change, if not identify the exact cause. One can follow initial debugger-less investigation to make process faster and more effective before starting formal debugging. In different situations, different methodologies prove more appropriate and as a professional coder, you should know them all. The following methods are utilized to debug without using a debugger:

Identifying regression and locate the cause

It happens that something worked fine till a certain build version and then boom! It did not. In such cases, a direct and common sense approach is to locate the cause and identify the changes that went into the victim build and then take corrective measures.

Delta debugging

The scientific method of debugging is to establish a hypothesis on why something does not work. Then, test the hypothesis and accept or reject it depending on the test outcome. This is how issues are usually located manually. Delta debugging in simple terms is automating this process.

Progressive unit testing or Saff-Squeeze

Kent Beck, co-father of XP, TDD and JUnit coined the term `Saff-Squeeze` and `Hit'em high, Hit'em low` to describe this methodology. The idea is said to be derived from American Football occurrence where the ball carrier is hit simultaneously by two players, one aiming upper body and other aiming lower body. Correspondingly, in this approach, a failing high-level unit test(high-tracker) is progressively replaced with more specific unit tests(low-trackers) until a test exists that directly identifies the problem.

Interesting, isn't it? Now let's discuss core-dumps, debug builds, debugger and the debugging process itself. These shall help you get a better picture of debugging process easily without getting entangled in unnecessary *Why* and *How* doubts.

Core-Dump or Memory-Dump files

Also referred as 'memory-dump' or just 'dump', a core-dump is a full-length snapshot of the memory (RAM) captured at a certain instant and stored to hard disk. A core-dump is generally collected when the program crashes i.e. terminates abnormally. In Linux, one may also force a normally running process to core-dump by sending signals like SIGABRT:

```
1.  $ kill -SIGABRT <process_id>
```

Man page of `signal` provides a list of signals that can cause a process to dump core. If you would like the process to dump core at a certain point in the code, you can do that by adding a call to `raise()` or `abort()` library functions. Core-dumps allow for offline debugging and are very useful in several situations. For example,

- Live debugging may not always be possible as customers may not allow access to their machines for different reasons. In such cases debugging with the help of logs and core-dumps is the only option available. Sometimes reproducing the bug requires certain setup or situations which cannot be created locally - One of the authors once came across a bug which was randomly being observed only in a certain machine and could not be reproduced anywhere else. All the attempts to reproduce on other machine did not result in success. For some data security reason, the customer would not allow access to machine for live debugging.
- Also, a client may not afford to have a busy system out of production for long. If some process crashes, it must be brought back to life at the earliest to continue operations. With core dump generated while the process crashed, what caused it could be analyzed separately.

It is important to understand that a core-dump just shows manifestation of a bug- the point where process cored may not be the location of bug. The actual cause may be somewhere else which in most cases, can be traced back to exploring the flow of execution.

To allow for cored-dump to be generated, you need to set different variable in different kind of shells indicating the allowed size for a core-dump. In bash, it can be done by setting *ulimit* to a certain value or *unlimited* like this:

```
1.  $ ulimit -c unlimited
```

To have core-dump collected in case of crash, make sure that-

- Process has permission to write to directory where core is to be dumped.
- There is no existing file with the same name.
- There is enough space on system.
- The specified directory must exist.

By default, the core dumps are generated in /home/<user>/core. However, this location can be changed. Not only that, even the naming pattern for the core can be specified. You wouldn't want every core to be named "core" else it may keep overwriting the previous dump every time. Use /proc/sys/kernel/core_pattern to provide all such settings. You may see manpage of core to see more details about this.

Also, a core-dump in itself is not sufficient for debugging as it presents the memory contents as a formatted series of lines indicating the memory locations and values recorded at each location in hexadecimal. For example, you may get information such as `executing instruction 0x54321 at 0x76543 resulted in dereferencing a assert failure`. However, being able to see the function name and line number would be more helpful and intuitive to the user. To do that, something called as symbol tables are used by debugger to map the instructions in hex to the code.

Symbol Table

Debugging is only possible if the binary you are trying to debug has debug symbols available with it. A debug build is a collection of binaries where all of the involved binaries have the debug symbols present in them. These debug symbols are represented as a symbol table.

Symbol table is a data structure used by a language translator where each identifier in the source code is associated with the information related to corresponding debug symbols. A debug symbol is the information which specifies that within an executable which portion of the machine code refers to which programming language construct. In other words, a symbol table is a mapping of an instruction with its variable name, function name or the line and file number information in the source code. Mostly symbol tables are implemented as hash tables. However, some of the implementations use linked lists and/or trees as well.

Symbol tables are not generated implicitly; we need to tell the compiler to generate them as a part of normal code compilation. For gcc, we need to pass "-g" option to generate a debug version of the binary or object file. A simple example is:

```
1.  $ gcc -g -o debugobj file.c
```

The object file "debugobj" thus created will have all the debug symbols which can be exploited by the debugger later on.

Debugger-Under the hood

The next section shall describe how to use a debugger; this portion is about how it works. You shall soon realize how wonderful a tool it is, if haven't already. We shall be using gdb (GNU debugger) on Unix/Linux as an example, arguably the most preferred and used debugging tool. The same principles apply to other debuggers and platforms as well.

What a debugger does is to let user access the memory of process being debugged. If you have been paying attention in your Operating Systems classes, you would find it awkward. You would know that a process reading or writing to the memory of another process is a serious security threat. So, how does debugger do it? Of course it can't if kernel doesn't allow it to do that. Debugger support comes built in with the Operating System kernel. Kernel can access the memory belonging to every process in the system. Also, when the process is not in running state, it can access information like value in registers. It helps to know where the process being debugged is stopped.

The debugger first needs to inform kernel that it is a debugger and is going to debug a certain process. The kernel can accept or deny this request. Also, the ability to read and set values from the memory space and registers of process being debugged is needed. Operating System provides way to do that. Linux provides `ptrace()` system call which is defined in `sys/ptrace.h`. If interested to go deep, you may explore more about ptrace() by yourself. The scope of this book doesn't cover it.

There are two ways of debugging a process:

Run the process from inside the debugger

In this method, the debugger is invoked first and from within the debugger, the process to be debugged (let's call it debugee) is invoked. The debugee gets invoked as a child process of the debugger and issues *PTRACE_TRACEME* to kernel telling that it wants to be debugged by its parent. Kernels allowing the same allows parent process debug the child.

Attach the existing process to the debugger

In this method, the process is already running and debugger needs to be made its parent in order to debug it. In order to do that, the debugger issues *PTRACE_ATTACH* to kernel. It tells kernel that for debugging, the calling process should be made parent of the process being called.

Therefore, in both cases, the debugger becomes the parent of the debugee process. At this point, the debugger presents us with debugee stopped so that we may make necessary preparations for debugging it. The preparations would mean setting up breakpoints and/or watchpoints, load symbol files and so on. Once ready, the debugee can be run and debugged. The stopping and running of debugee is done by Operating System using signals. Signals are beyond the scope of this chapter but it is recommended that you explore more about them.

The 'right' Debugging Process

The process of debugging is entirely different from that of writing code which is practiced often but rarely taught. After finding an issue, following the debug process properly can be quite helpful and less time consuming. This process can be explained as follows:

Checking if actually a bug

Not all bugs are obvious and not all reports are actually bugs. Therefore, first thing is to identify if the report is actually a bug. Sometimes the observed phenomenon is expected behavior while the product is being developed or it might be already fixed in new builds but the tests would be running on the older builds. Taking care of this step alone can save your precious hours.

Collect complete information

In software, a trigger somewhere can cause catastrophic something elsewhere. The relation between a process crash and something as simple as running a simple command or clicking somewhere might not always be straightforward. Devtest or customer may provide description like "*I did so and so and something weird happened*" but that would not happen when you try to do so. Get as much information as possible including the system condition, anything unusual that was done on system, steps to reproduce, logs and build information at the least.

Reproduce the issue

As the name suggests, it means replicating the mentioned undesired behavior in a controlled environment to identify the precise steps for reproducing the issue. This step is really important because it provides you the means to verify your fix. Failure to replicate the issue means failure to verify the fix which may potentially result in another bug.

Note: Although not very common, sometimes issues are peculiar to hardware or system configuration that can't be replicated locally would be exposed. In those cases, there is no definitive way but to rely on input received; judge and make the fix and provide it to them for testing.

Understand the issue and identify the root cause

As mentioned above, the relation between the trigger and the mishap might not be straightforward. Once you're able to reproduce the bug, understand what is actually happening and identify the root cause. Usually this step takes maximum time. Many times, to understand the issue, you need to understand the process.

If you aren't the author of the code, the previous chapter i.e. *Understanding the code* would be pretty helpful.

In this step debugger comes into picture as a tool that helps you look deep into what's going on inside the program. You may do core dump analysis, live or remote debugging depending upon situation. If it's a crash, debuggers can take you directly to the scene of aftermath i.e. where the crash happened but in case of logical errors, you'll need to explore more to see where things are going bonkers.

Making and verifying the fix

Making the fix usually is a fast step unless a huge change is needed. As you understand the issue and root cause, you automatically understand what needs to be done. After making the fix, you need to verify it and that is where the setup done for *Reproducing the issue* comes into picture. The fix should be verified thoroughly for both positive and negative test cases.

Making sure that the fix isn't another bug

Sometimes butterfly effect is observed in software i.e. a seemingly harmless change here can cause havoc elsewhere. It is therefore recommended that after making a change, not just the portion where modifications are done should be tested but other test cases should also be thoroughly covered to make sure that the fix is not resulting into another bug (or many).

The most important thing is to learn from the bug after fixing it. There might be other areas in program where similar conditions can arise. A good programmer also considers if the same thing could have been coded in a different manner to avoid issues.

Debugging with gdb

This section describes the usability aspect of gdb. If you have reached here, then you must be aware of the concepts of 'post- mortem' analysis and live debugging. Let us go through each of them in detail with apt examples to help you become a master of debugging.

Core analysis

Consider the following sample program:

```
1.  void func1(char *str) {
2.      char tempvar;
3.      printf("%s\n", str);
4.      str = 0;
5.      tempvar = *str;
6.      printf("%c\n", tempvar);
7.  }
8.
9.  int main(void) {
10.     char *temp = "tempstring";
11.
12.     func1(temp);
13.     return 0;
14. }
```

In line number 4, we are deliberately setting the character pointer "str" to NULL(o). In the next line (line number 5), we are dereferencing the same null pointer. This deference is going to cause the program to exit and dump a core file.

```
1.  $ gcc -g core.c
2.  $ ./a.out
3.  tempstring
4.  Segmentation fault: 11 (core dumped)
```

The above mentioned commands would generate a core file named a.out.core in this case. With all the pre-requisites achieved, let us see how to use gdb to pin point the exact root of this failure. Usage is pretty simple: gdb <binary with debug symbols> <core file>. Once you execute this command, you will land into gdb shell.

```
1.  $ gdb ./a.out ./a.out.core
2.  Core was generated by 'a.out'.
3.  Program terminated with signal 11, Segmentation fault.
4.  Reading symbols from /lib/libc.so.6...done.
5.  Loaded symbols for /lib/libc.so.6
6.  Reading symbols from /libexec/ld-elf.so.1...done.
7.  Loaded symbols for /libexec/ld-elf.so.1
8.  #0 0x08048553 in func1 (str=0x0) at core.c:5
9.  5 tempvar = *str;
10. (gdb)
```

The second line tells us that the program exited with signal 11 (Segmentation fault). This itself is a good start in understanding the crash. In this current state, gdb is waiting for the user input. The first command to try out on the gdb prompt is "backtrace" or simply "bt".

```
1.  (gdb) bt
2.  #0 0x08048553 in func1 (str=0x0) at core.c:5
3.  #1 0x0804859e in main () at core.c:12
```

This command gives us the details of the function call chain that caused this program to crash with the proper filename and line numbers. Now that you know the program exited while executing statement at line number 5, you can directly look at the source code to check what could be the possible reason for this behavior.You can also use "list" command to view the source code inside the gdb shell itself.

```
1.  (gdb) list
2.      void func1(char *str) {
3.          char tempvar;
4.          printf("%s\n", str);
5.          str = 0;
6.          tempvar = *str;
7.          printf("%c\n", tempvar);
8.      }
```

Statement at line number 5 is dereferencing a pointer and just by the sheer look at it tells us that segmentation fault can only come if the memory pointed to by variable "str" is not valid. To confirm this we can use "print" or "p" command which is used to examine the value of any variable inside gdb.

```
1.  (gdb) p str
2.  $2 = 0x0
```

Thus we know for sure that it is indeed a null deference which caused this program to exit. What if this pointer was passed as NULL from the caller of this function. To confirm this, we need to examine value of variable in the caller function and not in the current one. gdb provides command "frame" or "fr" to move between different functions in the call frame (frame numbers are already provided in the backtrace output).

```
1.  (gdb) fr 1
2.  #1 0x0804859e in main () at core.c:12
3.  12 func1(temp);
```

This command brings us to the place from where the function `func1()` was called. Once you are inside frame 1 which is "main()" in this case, you can again use "print" command to examine the values of the variables.

```
1.  (gdb) p temp
2.  $3 = 0x804862b "tempstring"
```

This clearly tells us that the argument had proper value before calling function `func1()`. So, we can confirm that the value got corrupted somewhere within the scope of `func1()` only.

See, it wasn't rocket science to debug and finally figure out the root cause for a defunct program that probably exited and dumped core at a customer's site.

Live debugging

Okay, so now you can do the 'post-mortem' of what has already crashed. How about an 'operation' of the process which is still alive? Using a debugger like gdb, we can check out the intrinsic details of an alive process.

If you are able to reproduce the issue by following certain steps, it is preferred to debug a live process and locate the cause as you can customize environment to pinpoint the root of the issue relatively quicker. That's because you get more freedom to play around with the process. As mentioned above, there are two ways to attaching a process to the debugger/gdb:

- **Running the process from inside the debugger**
  Get hold of the executable which already has the debug symbol information. Then, provide this path of the executable as a command line parameter to gdb:

```
1.  $ gdb ./a.out
2.  GNU gdb 6.1.1 [FreeBSD]
3.  Copyright 2004 Free Software Foundation, Inc.
4.  GDB is free software, covered by the GNU General Public
    License, and you are
5.  welcome to change it and/or distribute copies of it
    under certain conditions.
6.  Type "show copying" to see the conditions.
7.  There is absolutely no warranty for GDB. Type "show
    warranty" for details.
8.  This GDB was configured as "i386-marcel-freebsd"...
9.  (gdb)
```

Similar to the case of core dump analysis, this command will open a new shell waiting for the user's input. You need to start the program execution with gdb as its parent. This is achieved by command – "run" or "r". This command also takes the input command line parameters if your program accepts any.

```
1.  (gdb) r
2.  Starting program: /var/myprograms/a.out
```

- **Attach an already running process to the debugger**
  This can be done for daemons or processes which run for longer durations and you don't want to restart them. You may attach an existing process to the debugger as:

```
1.  $ gdb <path_to_executable> <pid_of_running_process>
```

This shall stop the process execution and take you to the gdb shell. At the shell, you need to execute the command "continue" or "c" to let the process continue its execution.

```
1.  (gdb) c
```

Once this initial setup is done, there is no difference in the way gdb handles the running or attached process. The beauty of a debugger lies in the control of the debugee program which it gives to the user. In terms of debugger terminology a "breakpoint" is a facility to stop the execution of the running program at will. gdb provides the facility to stop execution of a program at a given function call or a specific line of source code. Command used to place a breakpoint is - "breakpoint" or simply "b". You need to set the breakpoints before hitting "r" or "c", otherwise the program will start its execution and you won't be able to stop it again for setting the breakpoints.

Setting a breakpoint on the main() routine will stop the execution at the very first instruction and thus you can follow the program execution right from the beginning.

```
1.  (gdb) b main
2.  Breakpoint 1 at 0x804858c: file core.c, line 10.
```

Alternatively, you can set the breakpoint at a particular line number of source code by specifying the file name and line number in the following format – "b <filename>:<line number>":

```
1.  (gdb) b core.c:5
2.  Breakpoint 2 at 0x8048550: file core.c, line 5.
```

Once a breakpoint is set, and you hit "run" or "continue" the program will start executing till the breakpoint is hit. Once a breakpoint is hit, gdb again will offer you the shell and ask for your valuable inputs. You can always get to know about the breakpoints using command – "info breakpoint" or simply "info b".

Command "next" or "n" would clearly pass as the most used command in the gdb world. Once a breakpoint is hit and program execution is halted (in other words - waiting for user's input), you can use this command to execute one single statement at a time. Let us assume that gdb is waiting at the breakpoint set at main() which is line number 9 in our file core.c.

```
1.  (gdb) Breakpoint 1, main () at core.c:10
2.  10 char *temp = "tempstring";
3.  (gdb)
```

Here, gdb tells us that the next statement to be executed will be line number 10. Using "next" here will execute this statement and display the next statement to be executed:

```
1.  (gdb) n
2.  12 func1(temp);
```

Now, the next statement to be executed is the function call to func1(). Pressing next will take us to the next line of execution which is past this function call. But, what if we wanted to go inside this function and see what is happening there. gdb doesn't disappoint you. The command "step" or "s" is there for this purpose only.

```
1.  (gdb) s
2.  func1 (str=0x804862b "tempstring") at core.c:3
3.  3 printf("%s\n", str);
```

You can see that now we are inside func1() and the next statement to execute will be line number 3 i.e. the first line in this function. You can use "print" command to check the values of any variables like we did for the case of core dump analysis. To get out of this function simply keep on hitting "next" till the last statement of this function is executed or use "finish" or "fin" command to get out instantly.

You can use "continue" command to continue the execution of the program again. Once continued, the program will run till finish or till another breakpoint is hit.

gdb has a very rich and extensive command set to do anything you can ever think of, however the ones covered here should get you going quickly. You should refer to the gdb manual to know more. And yes, before we forget to mention – use `quit` or `q` to exit gdb.

Remote Debugging

Debugging is lot easier if you are able to map each instruction to the code. If the code is present on the same machine where it is being executed, the mapping is easy. However, when the code and the process are on different machines, remote debugging comes to our rescue.

Setting up remote debugging is a two-step process –

- Start gdbserver binary on the machine where the process to debug is actually running. gdbserver is a utility which allows remote connection on a TCP channel using standard gdb remote serial protocol. To start a gdbserver you need to specify a port along with the program and its command line arguments:

```
1.  $ gdbserver :1111 <executable path> <command line
    arguments>
```

To attach an already running program, use the following command:

```
1.  $ gdbserver :1111 -attach <pid of the program>
```

- On the remote/client machine where you have the source code for the same executable use the following command:

```
1.  $ gdb <executable path>
```

Once you are inside the gdb prompt, use the combination of ip address of the machine where gdbserver is running and the port specified in the gdbserver command to connect to remote gdbserver:

```
1.  (gdb) target remote <ip_address>:<port>
```

Now, you can use this gdb session as if you are doing live debugging on this local machine itself.

🔊 gdb is a very robust and reliable tool with very simple yet rich set of commands. Eventually you will find gdb becoming your best buddy in your professional life.

Other Debuggers

There are many flavors of debuggers available for your use apart from gdb. Intel debugger (IDB), WinDbg, dbx (debugger on solaris) are some examples. MicroSoft Visual Studio Debugger comes as an IDE with inbuilt source code viewer, compiler and debugger in one package. DDD is a GUI based debugger developed over gdb itself, in this you can view the source code along with command prompt. KDB is the kernel debugger for Unix related operating systems.

Apart from this, you can easily integrate gdb with popular code browser tools like emacs and vim to convert the simple command line experience into a GUI based one. With this you will be able to browse the source code and view the current line at execution with more clarity than what is visible through a command line based debugger.

Hope you had fun with debuggers, it's time to lay your hands on some of the popular program analysis and profiling tools. Next chapter shall equip with the knowledge that separates a stud from an average guy.

Chapter Review

- Debugging is a process of identifying and reducing the occurrences of bugs or issues in a software program.
- There are many ways to find the root cause of bugs – simple print statements, specialized debug tools, testing, finding the recent piece of code that could have brought this bug etc.
- GNU Debugger (gdb) is used as a standard tool to understand the debugging process in this chapter as this is most widely used debugger at present.
- Debugging can be classified as "post-mortem", live debugging and remote debugging.
- Debugging is only possible if we have a debug build which contains all the debug information in the form of symbol table. This information is generated by the compiler if the appropriate flag is passed to it.
- Debugger needs special privileges from Operating System to provide the functionality to stop and examine the intermittent state of a running program.
- A program can be debugged by starting it within the context of the debugger, or an already running program can be attached to a debugger.
- Gdb has a very simple yet very powerful set of commands. You can set breakpoints at specified location to make the debugger stop execution of the main program.
- Commonly used gdb commands are – "next", "step", "finish", "continue", "print", "bt" and "breakpoint".
- Core files are really helpful if a program crashed at customer end and it is hard to reproduce the issue locally.
- If you don't have the source code on the same machine where the executable to debug is running, use remote debugging with gdbserver.

Program analysis & Some More Important Tools

Your program runs. Good, but is it good enough? Probably not!

It is not enough to address only the functional aspects of a software. The non-functional aspects are as important. Should your program behave erratically during long runs or crash when exposed to real-world scenarios or end up losing precious data, it is far from being considered good software.

One disaster that you might have heard of as the first 'improved' online CAT exam organized by the company called Prometric. Various users faced troubles of different kind. The software might have met all functional requirements but failed to deliver when it was being accessed by a good number of people simultaneously. Mind you, the number of simultaneous users taking up the exam were nowhere close to what websites like Google and Facebook handle every second. Developers there take care of throughput, availability, scalability and so on while designing and developing software and thus they are able to cope up with sudden surge of connection requests.

Many times such problems are discovered only when the new use cases are identified which were not considered while creating the software.

Therefore, it happens that you find yourself in a situation to improve the existing code and make it devoid of such problems. However you cannot check on such aspects without using specialized tools and which is why this chapter discusses various specialized tools that would help in such situations. Also, one of the major causes of multiple issues i.e. immature handling of memory resource is covered at length.

Caution, this specialized knowledge can get you soon to the point where your peers would start applauding and hailing you.

Excited? On your marks, Get set, Go!

Memory Issues

Though your program might be showing expected outputs, it is quite possible that it's not free of memory issues. Memory related issues are among the difficult-most ones because they may usually not cause any observable problem and pass the basic tests but impact elsewhere in the program sometime causing 'randomly' observed issues which are not easy to reproduce and debug.

For example, suppose you forgot a corresponding *free* call in your code for one of the *malloc* statements. The program shall continue to run as expected until you've run out of memory and then very next call to *malloc* shall fail. The failed *malloc* may be at some other location in the program. Core-dump analysis shall give you the victim but may not help you in identifying the culprit i.e. where the root cause lies.

Caution: Though It's not proven formally but most developers shall agree to the fact that memory issues have deep respect for Murphy's laws - They love to throw surprises on Fridays and that also at major customer deployments which may translate into an 'adventurous' weekend for you. Therefore, you may want to consider these issues even more seriously to avoid such 'fun' weekends :)

Tackling memory issues needs specialized tools for the job. One of the most preferred and commonly used such tool is Valgrind. It was originally designed to be a free memory debugging tool for Linux but later it evolved to become a framework for developing various other tools as well. Valgrind is freely available under GPL license. You may freely download and install it. When you run valgrind, you need to specify the particular tool you would want to use. Next section focuses on **memcheck tool** and **leak-check** option. However, you may like to explore rest of them on your own.

Listed below are some of the most commonly observed memory related issues. You will see how Valgrind can be of help in tackling these issues:

- Invalid pointer usage
- Using uninitialized variables
- Memory leaks
- Multiple frees

Invalid Pointer Usage

Suppose you allocate some memory and try to access it beyond the allocated size like the following example:

```
1.  #include<stdio.h>
2.  #include<stdlib.h>
3.  int main(void) {
4.      int *array = malloc(5);
5.      array[6] = 1;
6.      return 0;
7.  }
```

Name this program as invalidPtr_test.c and simply compile it:

```
1.  $ gcc -g invalidPtr_test.c -o invalidPtr_test.out
```

This program runs successfully but it actually contains a memory corruption which however, may or may not get detected depending upon whether the overwritten location would be gettIng utilized elsewhere. Such discrepancies may elude programmer's eyes but Valgrind shall detect it. All that is required to be done is running the executable through Valgrind like this:

```
1.  $ valgrind --tool=memcheck invalidPtr_test.out
```

This shall give you an output like this:

```
1.  ==12004== Memcheck, a memory error detector
2.  ==12004== Copyright (C) 2002-2011, and GNU GPL'd, by
    Julian Seward et al.
3.  ==12004== Using Valgrind-3.7.0 and LibVEX; rerun with -
    h for copyright info
4.  ==12004== Command: ./invalidPtr_test.out
5.  ==12004==
6.  ==12004== Invalid write of size 4
7.  ==12004== at 0x400512: main (invalidPtr_test.c:6)
8.  ==12004== Address 0x51f1058 is not stack'd, malloc'd or
    (recently) free'd
9.  ==12004==
10. ==12004== ERROR SUMMARY: 1 errors from 1 contexts
    (suppressed: 4 from 1)
11. ==12004== malloc/free: in use at exit: 0 bytes in 0
    blocks.
12. ==12004== malloc/free: 1 allocs, 1 frees, 5 bytes
    allocated.
13. ==12004== For counts of detected errors, rerun with: -v
14. ==12004== All heap blocks were freed -- no leaks are
    possible.
```

Check line number 6 of the output, It clearly tells that there is an invalid write happening there. That is because the write is being done beyond the allocated range of 5 bytes. Valgrind even provides information that the issue is in *main* function at line number 6 in the file invalidPtr_test.c.

Note: Valgrind shall be able to show you the line numbers in the trace only if you have compiled your binary with "-g" option of gcc which adds debug symbols to it.

Using uninitialized variables

This example is pretty much self-explanatory. An uninitialized variable may contain any garbage value. Assigning the same to another variable or using for comparison would result in undefined behavior. As mentioned in first chapter, you should initialize all the variables.

Now consider an example to see how Valgrind can help you identify issue with uninitialized variables:

```
1.  #include<stdio.h>
2.
3.  int main() {
4.      int a;
5.      if (a == 100)
6.      {
7.          // Some operation
8.      }
9.      return 1;
10. }
```

Run the executable in Valgrind with the option *track-origins* set to *yes* to get details regarding uninitialized variables:

```
1.  $ valgrind --track-origins=yes
    ./uninitializedVar_test.out
```

Output generated by Valgrind will look something like this:

```
1.  ==12032== Memcheck, a memory error detector
2.  ==12032== Copyright (C) 2002-2011, and GNU GPL'd, by
    Julian Seward et al.
3.  ==12032== Using Valgrind-3.7.0 and LibVEX; rerun with
    -h for copyright info
4.  ==12032== Command: ./uninitializedVar_test.out
5.  ==12032==
6.  ==12032== Conditional jump or move depends on
    uninitialised value(s)
7.  ==12032==    at 0x400500: main (2.c:7)
8.  ==12032== Uninitialised value was created by a stack
    allocation
9.  ==12032==    at 0x4004F4: main (2.c:3)
10. ==12032==
11. ==12032==
12. ==12032== HEAP SUMMARY:
13. ==12032==   in use at exit: 0 bytes in 0 blocks
14. ==12032==   total heap usage: 0 allocs, 0 frees, 0 bytes
    allocated
```

As you can see 6th line onwards, Valgrind clearly identifies that the particular condition depends upon an uninitialized value. As seen in previous example; file, function and even line number details are mentioned.

Note: In some older versions of Valgrind, track-origins is not supported. For these versions, simply use Valgrind on the executable without specifying any option.

Memory leaks

Among memory related issues, memleaks are arguably the trickiest and most-difficult ones to locate and resolve. However, Valgrind makes life lot easier for developers while dealing with these.

Memory leaks or memleaks occur when a program acquires memory but after it is done using it, doesn't release the memory back to the operating system. In other words, when the number of malloc calls made are more than the corresponding free calls in a program, it contains memory leaks. Valgrind works on the same principle. To see what output you can get from Valgrind if the program contains a memory leak, consider the following code:

```
1.  #include<stdio.h>
2.
3.  int main() {
4.      int *x = malloc(10);
5.      return 1;
6.  }
```

Run the executable in Valgrind with the option *leak-check* set to *yes*:

```
1.  $ valgrind --tool=memcheck --leak-check=yes
    ./memleak_test.out
```

And the output generated for all the memleaks for the above example will look something like this:

```
1.  ==12146== Memcheck, a memory error detector
2.  ==12146== Copyright (C) 2002-2011, and GNU GPL'd, by
    Julian Seward et al.
3.  ==12146== Using Valgrind-3.7.0 and LibVEX; rerun with -
    h for copyright info
4.  ==12146== Command: ./memleak_test.out
5.  ==12146==
6.  ==12146== HEAP SUMMARY:
7.  ==12146== in use at exit: 10 bytes in 1 blocks
8.  ==12146== total heap usage: 1 allocs, 0 frees, 10 bytes
    allocated
9.  ==12146==
10. ==12146== 10 bytes in 1 blocks are definitely lost in
    loss record 1 of 1
11. ==12146== at 0x4C2B6CD: malloc (in
    /usr/lib/valgrind/vgpreload_memcheck-amd64-linux.so)
12. ==12146== by 0x400505: main (3.c:5)
13. ==12146==
14. ==12146== LEAK SUMMARY:
15. ==12146== definitely lost: 10 bytes in 1 blocks
16. ==12146== indirectly lost: 0 bytes in 0 blocks
17. ==12146== possibly lost: 0 bytes in 0 blocks
18. ==12146== still reachable: 0 bytes in 0 blocks
19. ==12146== suppressed: 0 bytes in 0 blocks
20. ==12146==
21. ==12146== For counts of detected and suppressed errors,
    rerun with: -v
22. ==12146== ERROR SUMMARY: 1 errors from 1 contexts
    (suppressed: 2 from 2)
```

You can see in the output that 10 bytes are definitely lost i.e. we have a memory leak where 10 bytes are not getting freed. Successive lines shows the function call trace which can help in identifying where the corresponding free should be done.

Using "--leak-check=yes" is usually sufficient. However, there might be cases where you would need more stringent checks. If you suspect that there are more memleaks than what are identified by "–leak-check=yes", then you may add "--show-reachable=yes" to your command. There shall be no difference in the output except that it may show even more cases where memory might not be freed.

Multiple frees

Calling multiple free on a pointer may lead to issues. If you are calling multiple free() corresponding to a malloc, valgrind shall let you know that you are calling an "Invalid free()". Of course, it shall provide a stack trace like in other cases.

Valgrind checks out of bounds for dynamically allocated memory only. Therefore, if you define an array inside your function but access an element beyond its size, valgrind shall not catch that.

Profiling and Performance Analysis

Okay, so you have written a piece of bug-less code that works - You supply input and within split of a second it provides expected output, Great! But can it do better? Can it be optimized to do the same stuff in lesser time? Are there any bottlenecks that may hamper its execution or slow it down in certain scenarios? When the program is composed of thousands of modules and millions of lines, which areas are consuming maximum time or should be focused on to get maximum improvement in performance?

A profiler is something that can help you in getting answers to these questions by providing a dynamic analysis of a program. Using which you can identify the point of concerns and take necessary measures for improvement. For example, one of the writers of this book was asked to optimize certain code which was taking pretty longer to execute and bringing down the performance of whole system. All he did was - used a profiler and found that a lot of time was going in writing logs. He identified that some logs that provide regular information could be optimized. Some of the logs that would be useful only for debugging could be changed to debug logs which need not be written during regular execution. Just by creating MACROS that would enable debug logs only if debug option is set and optimizing on the unnecessary logs, he was able to record appreciable improvement in performance.

Without further ado, let's check out gprof - the GNU profiling tool, which is probably the most-used one. First you need to compile and link the program with profiling option enabled. You can do this by simply specifying *-pg* option with the regular gcc command you would use. For example:

```
1.  $ gcc file1.c file2.c -pg
```

If you are compiling and linking in two separate commands, use *-pg* option in both. Now, you can just run the program as usual providing the required inputs. The program shall run, however you may notice that it would be running a bit slower which is fine. The slowness would be introduced by the profiler collecting the necessary data.

Note: The profiler shall collect data only about the executed portion of code. The areas not visited or functions not executed for a certain type of input shall not be profiled.

Before exiting, the program shall dump the collected information to a file titled *gmon.out*. The existing *gmon.out* file, if any, would be overwritten. So, make sure to rename the file if you are planning to run the profiler again with different inputs. Also, *gmon.out* shall be written to the present-working-directory of the executable when it exits.

After you have got the gmon.out file, use gprof to interpret it like this:

```
1.  $ gprof <options> <executable> gmon.out >
    <output_file>
```

gprof provides various options to selectively include or exclude certain functions and filter your result. You may check those on the man page of gprof.

To understand gprof in action, let us write down a sample C program containing a few for-loops just to make it interesting. We need to have a few programming statement that takes some time to execute so that we can get some real data at our disposal and provide the same to gprof as an input.

```
1.  #include<stdio.h>
2.
3.  void fun() {
4.      return;
5.  }
6.
7.  void fun1() {
8.      int j = 0;
9.      for (j = 0; j < 0xffffabcd; j++);
10.          fun();
11.      return;
12. }
13.
14. void fun2() {
15.     int k = 0;
16.     for (k = 0; k < 0xffffffff; k++)
17.         fprintf(stderr, "Having fun?");
18.     return;
19. }
20.
21. int main() {
22.     int i = 0;
23.     for (i = 0; i < 0xffffffff; i++)
24.         fun1();
25.     fun2();
26.     return 1;
27. }
```

Let us name it *gprof_test.c*. Now you can compile this file with profiling option, and then run it to collect *gmon.out* file. Once you have *gmon. out* file, use *gprof* to use this file to generate the profiling information. Following is the list of commands which you need to execute to achieve the same:

```
1.  $ gcc gprof_test.c -o gprof_test -pg
2.  $ ./gprof_test
3.  $ gprof gprof_test gmon.out > outfile
```

The 'outfile' thus generated consists of two parts namely **Flat Profile** and **Call Graph**. Following is a snippet of Flat profile which displays the total time spent in executing each function:

```
1.  Each sample counts as 0.01 seconds.
2.  %     cumulative    self              self     total
3.  time    seconds    seconds    calls   us/call  us/call name
4.  75.74     0.02       0.02   16711425    0.00     0.00 fun
5.  25.25     0.02       0.01      255     19.80    79.21 fun1
6.   0.00     0.02       0.00        1      0.00     0.00 fun2
```

Call graph shows how much time was spent within each function and its children. Following is a snippet of the same:

```
1.  granularity: each sample hit covers 2 byte(s) for 49.51% of
    0.02 seconds
2.
3.  index % time    self    children    called         name
4.                  0.01      0.02      255/255         main [2]
5.  [1]     100.0   0.01      0.02      255             fun1 [1]
6.                  0.02      0.00 16711425/16711425 fun [3]
7.  -------------------------------------------------------
8.                                               <spontaneous>
9.  [2]     100.0   0.00      0.02                      main [2]
10.                 0.01      0.02      255/255         fun1 [1]
11.                 0.00      0.00       1/1            fun2 [4]
12. -------------------------------------------------------
13.                 0.02      0.00 16711425/16711425 fun1 [1]
14. [3]      75.0   0.02      0.00 16711425             fun [3]
15. -------------------------------------------------------
16.                 0.00      0.00       1/1            main [2]
17. [4]       0.0   0.00      0.00       1              fun2 [4]
18. -------------------------------------------------------
```

Each entry in this table has primary line starting with index number in square brackets like [1]. This line specifies the function which this entry corresponds to. Preceding lines describe callers of this function and following lines shows the functions it calls.

%time specifies the portion of time spent in this function or its subroutines.

self shows the total amount of time spent in that function. It would be identical to *seconds* field for the function in Flat profile.

children shows the time spent in the subroutines called by this function

called shows the number of times the function was called. The two numbers specifies the number of times function was called from a certain function and the total number of non-recursive calls to a function made from all its callers.

name as obvious, shows the name of the function

Note: There are number of tools like kprof, Gprf2Dot and so on that can be used to produce a graph kind of visualization from the gprof output which is much easier to understand.

Coverage analysis

Code coverage helps you test your code and identify whether all the cases are covered during testing. The analysis of the same is helpful in creating more efficient and faster running programs and also to discover if test cases covered all of the code or not. If some cases are not covered, it may result in horrendous errors escaping to the production environment and causing catastrophic damage.

Like profiling, coverage can also help in easily identifying the areas where the optimization efforts shall affect most to achieve better performance. A function taking long is definitely a bottleneck but a function that may not be taking very long in individual calls and if it is getting called numerous times and hence collectively consuming a good chunk of time should definitely be looked into. May be its possible to reduce the number of calls to this function saving the time that was getting consumed in context switching. For example, during the assignment mentioned earlier, the same developer used gcov and identified that a small function which was doing a trivial task was getting called thousands of times. As the function was simple and small, it could not harm to make it inline. Well, that's exactly what he did and it contributed to a noticeable appreciation in the performance of the system.

To use gcov (GNU test coverage tool), first you need to compile the code with *-ftest-coverage* and *-fprofile-arcs* options. *ftest-coverage* option adds the instruction to count the number of times individual lines are executed while *fprofile-arcs* tells compiler to add instrumentation code for each branch of the program e.g. conditional statements like 'if', 'switch' etc. The program can then be run to produce the coverage data which can be analyzed to see how many times each line was executed and what portions of the code are not traversed during test run. Following program can be used to demonstrate the same:

```
1.   #include<stdio.h>
2.
3.   int main() {
4.       int i = 0;
5.       while(i < 20) {
6.           if (i % 2 == 0)
7.               printf("%d: divisible by 2\n", i);
8.           else if(i % 3 == 0)
9.               printf("%d: divisible not by 2 but 3\n",
     i);
10.          else if(i % 5 == 0)
11.              printf("%d: divisible not by 2 or 3 but by
     5\n", i);
12.
13.          if (i%20)
14.              printf("%d: divisible by 20\n", i);
15.      }
16.      return 1;
17. }
```

Name this file as *gcov_test.c* and then compile the code with following options. Finally run it to capture the coverage data and then analyze it as follows:

```
1.   $ gcc -o gcov_test -fprofile-arcs -ftest-coverage
     gcov_test.c
2.   $ ./gcov_test
```

Running this "gcov_test" binary shall yield the following output:

```
1.   2: divisible by 2
2.   3: divisible not by 2 but 3
3.   4: divisible by 2
4.   5: divisible not by 2 or 3 but by 5
```

```
5. 6: divisible by 2
6. 8: divisible by 2
7. 9: divisible not by 2 but 3
8. 10: divisible by 2
9. 12: divisible by 2
10. 14: divisible by 2
11. 15: divisible not by 2 but 3
12. 16: divisible by 2
13. 18: divisible by 2
```

Now run the gcov tool over this binary:

```
1.    $ gcov gcov_test.c
2.    File 'gcov_test.c'
3.    Lines executed:92.31% of 13
4.    gcov_test.c:creating 'gcov_test.c.gcov'
```

Here is the generated annotated version of the source code with extension .gcov.

```
1.         -:   0:Source:gcov_test.c
2.         -:   0:Graph:gcov_test.gcno
3.         -:   0:Data:gcov_test.gcda
4.         -:   0:Runs:1
5.         -:   0:Programs:1
6.         -:   1:#include<stdio.h>
7.         -:   2:
8.         1:   3:int main() {
9.         1:   4:    int i = 1;
10.       21:   5:    while(i < 20) {
11.       19:   6:        if (i % 2 == 0)
12.        9:   7:            printf("%d :divisible by 2\n", i);
13.       10:   8:        else if(i % 3 == 0)
14.        3:   9:            printf("%d: divisible not by 2 but
      3\n", i);
15.        7:  10:        else if(i % 5 == 0)
16.        1:  11:            printf("%d: divisible not by 2 or 3 but
      by 5\n", i);
17.        -:  12:
18.       19:  13:        if (i%20 == 0)
19. #####:  14:            printf("%d: divisible by 20\n", i);
20.       19:  15:            i++;
21.        -:  16:    }
22.        1:  17:    return 1;
23.        -:  18:}
```

This output shows the count for each line telling how many times it was executed in the process. You can see that the line count i.e. the number of times a line was executed is displayed at the start of each line. Lines which were not executed at all are marked with '#####'. To see the portions of the program that were not executed, you may just do a grep for '#####'.

🔊 Tools like grpof, gcov and Valgrind need not be run on daily basis and by everyone. In fact there is a specific team in every organization commonly known as system test team which keeps on doing such analysis from time to time.

This marks the end of our tryst with program analysis and profiling tools and how you can utilize this knowledge to save yourself in trying times. Next chapter focusses on an important aspect of software life cycle which one generally tends to overlook – software testing. However, try not to bypass it this time. Continue reading to find out, why!

Chapter Review

- Memory related issues are hard to find as they popup randomly and that too under certain conditions. They make their presence felt rarely, so you need to be proactive in finding potential memory related issues.

- Valgrind is a free memory analysis and debugging tool which can catch a variety of issues e.g. Invalid pointer usage, uninitialized variables, memory leaks, multiple free etc.

- For Valgrind to work properly and fetch the line number and file name information, the executable should be compiled with debug symbols.

- Memory leaks are caused when a program fails to release the memory it has acquired during program execution earlier. Valgrind helps in capturing potential memory leaks by providing "--leak-check=yes" option at the command line.

- Gprof is a tool that aids in analyzing the run time performance of a software program. One such analysis that is highly useful is generating a timing analysis consisting of the time spent in each and every function call.

- Gcov is a code coverage tools which analyzes the source code to see if all the cases are covered or not. If the report is showing zero coverage for a particular code snippet, either your tests are not sufficient or your code is simply a dead piece of code.

- It is not mandatory to run these tools on daily basis but it surely improves the quality of product if such analysis is done once in a while.

7

Software Testing

So you write a program, successfully compile and link it. Now you have a binary in hand. What's the next obvious step? Of course, to run it and see if it is correctly doing what it is expected to. Well that pretty much sums up in one line what this chapter is about.

Before you say, "Thats it?", please be reminded that the straightforward looking things are not always so straightforward, not in corporate world at least, which is why we have a full chapter devoted for the same.

In the traditional waterfall model, the testing team would pitch in after the development work was concluded. However, this is not the scenario anymore. In agile environments, testing is an integral part of the complete process and consists of varied phases where developers and testers are constantly engaged.

In this chapter, we are NOT going to bore you to core with the terms like black-box testing, white-box testing, grey-box testing and what not. You will instead, understand its significance and get a sneak peek into how it is done in the industry. Let's start with the significance portion.

Significance of testing

In college, you would take your "adding two integers" program and run it couple of times for some small value integers and drool over when it would magically produce the correct output. Perhaps, you wouldn't care much if it goes berserk upon running with huge values, strings or floats. After all, the description said integers, isn't it? You also would not care much about your program being portable to other architectures or its memory consumption or how scalable it is, right? Well, your program might be a working one but one meeting high quality standards? Most probably not!

When your programs are doing much more complex operations, being run in different environments and on different architectures and most importantly, earning bread and butter for the organization, the software to go out in the market has to be of high quality. In simple words, we can say that high quality means:

- Having all the desirable features working well.
- Being good, least confusing and most easy to use for the targeted user.
- It should not only work but must work as expected in the known and expected regular real world scenarios.
- It might have bugs but no 'show-stopper' ones.

To ensure this is the responsibility of everyone involved in the software development process. As you'll read later in the chapter, even customers get directly involved here which should give you a hint about the significance of Quality Assurance.

Manual and Automated testing

You would already have done manual testing while writing programs. When the programs are smaller and less complex, doing manual testing might look more straightforward, faster and convenient for the user. However, when the software is huge and complex and the same tests are required to be performed again, and again, and again, it makes more sense to automate some portions of the process, if not the complete process. To do the same, the initial efforts may be a bit time consuming but they are far more rewarding in the long run which is why companies prefer automating the testing process as much as possible.

Automating the testing process helps in many ways such as -

- Lesser chances of missing out test cases.
- Create once, use several times.
- Can enforce and drive clean design decisions.
- Saves and helps in better management of time and resources as test cases can be started and left to run while you may be taking rest or going for lunch or doing something more productive.
- Reduces chances of human errors e.g. making typos and having to type commands again.
- While refactoring i.e. restructuring the internal body of the code without affecting its external behavior, automated tests provide a sure-shot way to verify the changes.
- Reduces person dependency which helps in maintaining the same quality of the task regardless of who performs it.
- Writing test cases is challenging and not boring like manual one, and therefore provides ample opportunities to learn and develop technical and analytical skills.

To automate the testing, setups created are also termed as **test-beds** or **test-suites** which constitute machine(s), configurations that must be deployed to these machines and test plans that should be executed.

Most of the automation is done using scripts and therefore having good knowledge of scripting languages is required in this field. There are also numerous tools available in the market that can be utilized for automation testing. Some of the widely used examples are *HP Quick Test Professional, IBM Rational Functional Tester, SilkTest, Testing Anywhere, TestComplete, Selenium,* and *Visual Studio Test Professional.* Some companies prefer to create their tools in house and hire many engineering teams specifically for the job.

The following image is a snapshot of *Testing Anywhere* Client. Don't be scared of all the numbers, it's actually pretty easy and self-explanatory.

Testing Anywhere Client

File Edit View Tools Help Feedback

Record ● | New Test 🔍 | New Test 🔍 | New Run Suite ⟳

Tethys Solutions
Testing Anywhere

• Web Recorder
• Object Recorder
• BOL $18,513.32+

Run ⟳ | Edit ✎ | Delete 🗑 | Upload ↑ | Create EXE 📄 | Actions ▶

Folders

🗀 Testing Anywhere
  ⊞ 🗀 My Docs
  ⊞ 🗀 My Projects
  ⊞ 🗀 My Run Suites

File Name	Type	Repeat	Status	Last Run Time
ReliabilityTesting_041109.rnst	Run Suite	Repeat until I stop	4 Complete, 0 Failed	04/11/2009 10:40:19 PM
Bug3155.rnst	Run Suite	Repeat for 10 00:00 time	2 Complete, 0 Failed	06/12/2009 5:56:00 PM
Reliability.rnst	Run Suite	Repeat for 10 00:00 time	192 Complete, 0 Failed	07/12/2009 5:22:40 AM
ReliabilityTest-061109.rnst	Run Suite	Repeat for 05:00:00 time	7 Complete, 0 Failed	07/11/2009 12:27:57 AM
Bug3113.rnst	Run Suite	Repeat 3 times	3 Complete, 0 Failed	06/12/2009 6:38:20 PM
11.rnst	Run Suite	Do Not Repeat		
12Nov.rnst	Run Suite	Do Not Repeat	Complete	12/11/2009 12:24:54 PM
17thtest.rnst	Run Suite	Do Not Repeat	Complete	17/11/2009 11:26:10 AM
2ndNov.rnst	Run Suite	Do Not Repeat	Complete	02/11/2009 7:45:14 PM
3NovDatabase.rnst	Run Suite	Do Not Repeat	Complete	03/11/2009 6:42:34 PM
3NovMsgBox.rnst	Run Suite	Do Not Repeat		
AllDisabledTests.rnst	Run Suite	Do Not Repeat	Complete	27/11/2009 7:10:29 PM
Bug#2871.rnst	Run Suite	Do Not Repeat	Complete	17/11/2009 12:02:06 PM

Visual Logs | Properties | Schedule

■ Pass (3)
■ Fail - Error (5)
■ Fail - Checkpoint
■ Not Run (4)
■ Disabled

View Log □ | Delete 🗑

No ▼	Logs	Start Time	End Time	Status
4	ReliabilityTesting_041...	04/11/2009 10:18:5...	04/11/2009 10:40:19...	Total:12 Passed:3 Failed:5 Disabled:0...
3	ReliabilityTesting_041...	04/11/2009 9:31:48 ...	04/11/2009 10:13:51...	Total:23 Passed:17 Failed:6 Disabled:0...
2	ReliabilityTesting_041...	04/11/2009 8:44:45 ...	04/11/2009 9:26:48 ...	Total:23 Passed:17 Failed:6 Disabled:0...
1	ReliabilityTesting_041...	04/11/2009 7:57:30 ...	04/11/2009 8:39:45 ...	Total:23 Passed:17 Failed:6 Disabled:0...

Report Dashboard

Server Communication

Report Designer

Workflow Designer

41 run suite(s) | Username: Admin (Offline) | \Testing Anywhere\My Run Suites

Enabled by SMART Automation Technology®

Test Plans

Whether the testing is done manually or using automation, the very first thing to do is to create a *test plan*. A test plan defines what to test; it is a document that describes an ordered approach to test the software. The test plans are charted out according to the requirements. Creating good test plans is challenging as it requires thorough understanding of the product, assessing the software from various use case perspectives and how it integrates and works with other software.

The test plans are reviewed and agreed upon by both development and QA teams. While designing new products, even the other stakeholders like engineering product managers may get involved into reviewing these plans to ascertain that they cover all the design specifications.

As the bugs continue to get identified and resolved or new enhancements are added, corresponding test cases continue to get added to the test plan. Different test plans are created aiming different kinds of testing as described in the following sections.

Testing Levels

To ensure high-quality, the software has to go through various levels of testing during the process. Each level aims at testing the software from a different perspective. Here is brief description of each of these levels:

Unit testing

This is also known as the component testing and is performed by the developer as and when the changes are done for the software. The motto of this is to verify the functionality of the code added or modified and eliminate identified errors before passing the software to QA. It is usually done at functional level in procedural environments and at class level in object oriented environments.

Integration testing

Different teams work to build different components of the system which are verified to be individually working as expected in unit tests. Next, it is required to verify the interfaces between these components against software design. Integration testing seeks to expose issues in these interfaces and interaction between the components.

Note: It may be noted that components may use stubs while doing unit testing and are replaced by actual components at integration.

System testing

After all the components are put together to constitute a system, it is tested as a whole to ensure that it works as expected and also does not affect the operating environment in any way it is not supposed to. For example, consuming excessive amount of resources or corrupt the memory or stack.

Acceptance testing

This is usually performed by the user or customer to determine that the product actually meets the specified requirements and performs as desired in the kind of environments it is supposed to.

The testing performed at each of these levels is iterative and is constantly performed as software keeps getting modified to fix issues or add new features or enhancements. This covers the testing levels only briefly. However, at each or certain level different types of testing may be performed. Bear with us for now if this sounds confusing; it won't after you would have gone through the following section. So move on, shall we?

Testing types

Software is assessed in terms of various factors among which Reliability, Efficiency, Functionality, Usability, and Maintainability are the main ones. Different types of testing are designed to validate one or more of these. These are further classified into *functional* and *non-functional* testing.

In case you are not aware of or have forgotten these terms, *Functional Testing* focuses on testing the functions that code is expected to perform while the *Non-Functional Testing* focuses on testing other aspects like scalability, performance, security, and so on. Let's check out some majorly used testing types. Now you should be able to easily figure out which are functional and non-functional ones.

Ad-hoc Testing

This is an informal testing to perform random tests without referring to documentation or test cases.

Ageing testing

This is performed by running software normally for long durations like weeks or months, and observing if its behavior gets affected over continuous usage for long durations.

Installation testing

May times development and testing is done is virtual environments. However, the results on actual hardware are sometimes bit different. This is to ensure that the software gets installed properly and actually works on hardware of the kind deployed by customer.

Smoke testing

Smoke test is performed to verify that the critical functionalities of the program are working fine. It is done before performing detailed functional or regression tests to avoid wasting time on a 'broken' build. This is done using the test cases written to cover only the fundamental or most critical functionality of the software.

Sanity testing

Sanity test is performed to ascertain that new changes introduced as bug fixes or enhancements are working as expected and do not introduce any new issues. This does not mean thoroughly testing the new changes but at least check that for basic test cases it is working. If it's not, there is no point performing complex test cases. It may require adding new test cases.

Note: Smoke and Sanity test seem very similar and are therefore mostly misunderstood with one-another. In organizations, these terms are used interchangeably as sometimes both are performed together.

Regression testing

This is a thorough testing that is run with numerous test cases which have been added during the course of software development to ascertain that no regressions i.e. previously fixed issues have returned when new changes meet previously made ones.

Performance testing

This is performed to check attributes like stability, reliability and availability. It could be mainly divided into the following-

- **Stress testing**: This is done to check how much load the software can take, will it cause the system to hog, crash or slow down. The software is built to pass the test for loads much higher than it would normally be subject to or claimed by the company.
- **Endurance or Soak testing** - This is done to check if the software can take expected load continuously for weeks or months. Usually issues like memory leaks are discovered in this process.
- **Scalability testing** - As the name suggests, it is done to see how the software withstands scalability. The scalability may be in terms of increase in number of users, connections, transactions, size of database or even hardware.

Destructive testing

It is attempted to result in software or a portion of it to fail. Some also relate it to resilience testing i.e. supplying invalid or unexpected inputs and see how the software withholds. It is also called *failure testing* and mainly focuses on checking robustness of the system.

Usability testing

Mainly concerned with usage of the software to see if it is easy to use and understand by the supposed user.

Security testing

This is done to check how secure the software is and if it can thwart the hacking attacks aimed at system intrusion. It is mostly done on authentication, confidentiality and data integrity aspects of the software.

This should have clarified the term 'testing types' for you. The ones mentioned here are the most commonly used ones. There are many more testing types which have evolved along with the complexity of software and are more specific to certain scenarios. We would not like to overwhelm you with all of them and therefore covering those all is kept out of the scope of this book. You may easily learn about the ones that matter to you on the job or may explore more on your own.

Process followed in the corporate world

Okay, let's get to the exciting topic of what process is followed in the corporate world. It is pretty obvious that the development and testing keeps going on in a vicious circle. Within the organizations, there are certain practices followed to ensure that everything happens in a smooth and streamlined manner to avoid unpleasant surprises. These practices may vary a bit from one organization to another but the underlying concept remains the same. Here is an overview of how things are done:

- As a developer, before submitting your code changes to the software repository, it is advisable to put the built package through an automated test-suite that runs basic test cases for that package to ensure that the changes do not reopen a closed bug.

  Now a days in many corporations, a browser based interface is exposed where the developer can provide the location of his/her build package. The system puts it in the queue and as the request is dequeued, it picks the package from specified location and runs tests over it. The test results and the log location is mailed to the developer once the test is completed.

- At a specified time every day/night, the latest code is built into complete system and automated tests are run which tests the fundamental functionalities for overall system. In case issues are discovered, all those who recently made changes are notified and the details of failures are mailed to them.

  Note: In some of the companies, a build followed by automated running of basic tests gets auto-triggered as soon as changes are submitted to repository. This is fine for the systems that do not consume lot of resources. For the systems that do take hours together for the process, these steps are run at stipulated intervals e.g. every night.

- Depending upon the corporation, type of project and other factors, test cycles are planned. A test cycle refers to running all of the test cases on a certain build to verify the basic functionality, new changes and identify bugs. Usually, regular test cycles covering thorough test cases are run every week or even at shorter intervals in some companies.

- Performance or ageing test cycles that can run for days or weeks altogether are planned for longer intervals like once a fortnight or month.

  Note: Various test cycles may also be dependent upon each other. e.g. performance test cycle may be run only after Regression test cycle because if test cases are already failing, what is the point in putting it through a performance test.

- While creating the exact real world scenarios in a lab may not be possible because the real world is dynamic and many unforeseen conditions keep arising, a close to real-world scenario is emulated in the lab and tests are run on that to ascertain that software shall perform well once out. Such tests are more specifically performed on the builds that shall go to the customers whether as a beta build or production build.

This marks the end of an interesting discussion of a vital phase of software development without which any product is incomplete. We will see you in next chapter with a discussion on version control process and the tools to accomplish the same.

Chapter Review

- Maintaining quality of any software product is of essence because customers always expect it to work perfectly not only under the favorable conditions but also under unseen circumstances.
- Manually testing a small piece of software might be just fine, but as the system grows complex it is always not feasible to test it out manually. An automated system needs to be in place which leaves little chances for human errors.
- Scripts play an important role in automating a test system also known as "test-bed" or "test-suite". A test plan is an elaborate document on the test cases to be run along with the different environments under which testing needs to be performed.
- There are several different phases involved in a testing cycle – unit testing, integration testing, system testing and acceptance testing.
- Sanity testing is a form of testing where basic tests are performed upon addition of new features so as to ensure that older functionality is not broken.
- Regression testing is a thorough form of testing performed at regular intervals to ensure that not even a single feature or functionality is broken.
- Time to time testing team keeps on doing performance testing so as to evaluate the performance of the current system under various conditions specifically system under stress.
- Security testing is another important aspect of software testing where security related aspects are tested e.g. how the system will behave if someone tries to hack the system.
- Development and testing goes hand in hand and highly inter-dependent on each other. What follows in a professional environment is driven by the specific requirements of their own software product.

Version Control System

You have been working on a big assignment and in the process you have added some hundreds of lines of code spread across multiple files. All of a sudden your manager rushes with a high priority task and asks you to work on it. No problem! You would create a backup of the long term assignment somewhere with some description or timestamp as a part of the back-up directory. Now you happily finish the new high priority task.

Being a hard worker, you accomplish the job right on time and deploy the changes. All's well yet but Managers are, ahem! smart creatures. Next he/she tells you that you did great job, but this enhancement should also be the part of the long term work you are already working on. Now, it becomes bit tricky to imbibe all the new changes to your original task if the same files are affected. If you have a razor-sharp memory and are good at book-keeping tools, life would be easier and you will pick up your previous code and simply merge the currently deployed code with it. The task becomes really painful and time consuming if changes spans across multiple files and folders. It may appear like Deja-vu if many ad-hoc requests came in between and you could return to the original task only after weeks.

That's exactly what happens when you are working in a collaborative environment with several co-workers all sharing the same deployed code. On a day-to-day basis your changes will be clashing with that of someone else. You cannot just happily replace someone else's code, god knows what it may break. Moreover, suppose a change was a bad one and needs to be removed but while it was discovered, others added/modified some more logic there.

To avoid all this book-keeping and automating the process of code deployment rather than relying on developer's memory, professional firms use Version Control tools. If your reply to any of the following statements is yes, then you definitely need a version control system:

- Do you need to maintain multiple copies of the same code?
- Do you forget where the backup code and other files were stashed?
- You made some code change and later realize that you made a mistake and need to reverse the changes?
- You need to compute the difference between the multiple versions of your code?
- Want to identify which/whose changes caused/fixed an issue?
- Want to simply experiment on some piece of code without disturbing the already working code?
- You want to know who coded what? (Yes, you can find out if a person is actually a good coder or just a self-bragging idiot)
- You want to lock a particular file from being changed by anyone or you simply want to keep a track of developers modifying a particular file?

Basics of a version control system

Version control also known as source control or revision control, is the process of management of versions, revisions or changes done to a file or document. The core of any version control system lies in the concept of repository. A repository can be thought of as a collection of all the source files maintained in a particular directory structure. A simple repository of a code base may look like:

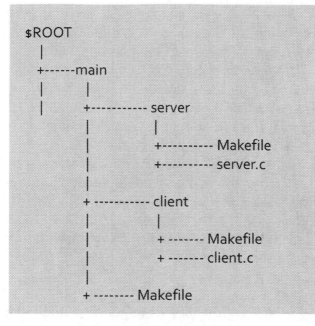

```
$ROOT
  |
  +------main
  |       |
  |       |
  |       +----------- server
  |       |            |
  |       |            +---------- Makefile
  |       |            +---------- server.c
  |       |
  |       + ----------- client
  |       |             |
  |       |             + ------- Makefile
  |       |             + ------- client.c
  |       |
  |       + -------- Makefile
```

Fig 8.1 Version control repository

The structure of a repository as depicted in Fig. 8.1 is easy to understand. It is a simple replica of the physical directory structure of code base with an addition of two more parent directories – root and main. We will come back to those later, till then assume it is nothing more than a convention to have them in the directory structure.

This repository acts as main depot for all future references for all users. You are not allowed to directly modify the repository. However, you can create a copy of this repository locally on your work station (we will call it "**workspace**"), do you changes locally without the intervention of others and then once you are done, submit your changes to the main repository. Once you submit your changes, they are available to all other co-workers who simply need to refresh their local copy to get your changes merged with their work space. So, you see there is no need for you to keep a track of other's code changes, you can always sync up your local copy to the latest version of repository and voila! You have the latest code changes of all your team mates.

Whenever a change is made to a file in the repository the version or revision number is also incremented. For example in the beginning the repository will be having revision 1 for file *server.c*, but as soon as some developer submits his changes, the version number changes to 2 and so on. This forms the core concept of versioning or maintaining revision history for any repository. With this revision history in place, developers can easily find the code changes done between two particular revisions. Usually developers submit multiple files as a part of their assignment; this information is also maintained in the form of 'changelists'. One 'submit' operation corresponds to one changelist which is assigned a unique key or hash for future references. Changelist contains information regarding the submit operation like the time, user information, list of files with their revisions etc.

Any version control system provides three basic functionalities: *parallelism, tagging* and *reversibility*.

Parallelism

We all work in a collaborative environment together and everyone should be able to work independently. Version control tools bring this parallelism by allowing anyone to maintain a local copy of the source code in their workstations which can be modified independently. As these changes are local to that particular user, it has no side effect on other users. With this concept of localization, multiple developers are able to perform changes simultaneously and no one is blocked because of other.

Tagging

Once a developer is done with making and testing his code changes, he needs to submit those to the main repository. It is only after the submit step, that the code changes are visible to others; otherwise everyone will die an agonic death tracking local and lost changes. When someone submits the changes, version control tools will maintain some meta-data for book-keeping. Date, time and developer information is almost always tagged along with any submit operation. Additionally, the users can and do specify a small description explaining why this change was needed and/or the mention of the bug this change is supposed to fix. This annotation is really helpful in future if someone wants to track down a code change for whatsoever reason.

Reversibility

Once a change is submitted to the main repository and later on the developer realizes that he made some mistake, he can always use the capability of version control tools to revert his changes. It is also possible to sync your local copy to a previously known good state of the repository if the latest one has some issues. Once the current state of repository is restored to a good state, you can again sync to the latest version. Thus reversibility feature allows all the developers to continue working on their local change even if latest repository is broken by some change done by an individual.

The repository itself can be classified into two different categories based on where the data resides.

Centralized

Centralized systems maintain the repository in a central server, something like a database. Server maintains all the files, directory structure and the meta-data for each submit and everything else that helps in maintaining file history. Normally the server will be a high performance and dedicated machine. This will be the only place where full historical data is maintained and all users will connect to this server and create a copy of files. If someone wants to submit their change, they will again have to connect to this server and then it will be reflected in the main repository and will be visible to all other users at the very same moment. Popular examples are ClearCase and Perforce.

Distributed

In distributed version control systems, the entire version/revision history is maintained on user's local machine somewhere under a hidden folder inside his local copy of source code. The information is in highly compressed format so having this information on local disk is not that much of a concern. As soon as you submit your changes, they get added as the meta-data in the hidden folder itself. You can simply copy the full folder along with the hidden folder to create multiple copies of the same project – this is what makes it a distributed system. The version control tool knows how to treat the changes separately for both the copies. So, effectively every user has his own repository locally which makes submit and sync operation real fast and that too without connecting to any external server. Usually, developers maintain a "master" copy in some external server so that others can share the same repository. CVS, Git and Mercurial are example of distributed version control systems.

Using version control system

This section describes the basic operations that need to be understood while working with any version control system. Although there are several flavors of version control tools, the basic terminology and usage remains fairly similar. Throughout this section we will discuss about the operations in general, for specific commands please refer to the manual provided with the tool you are using.

Add

A repository is empty when it is freshly created. You need to add files and folders to it using the "add" operation. This will create the first version / revision in the repository for the file / folder you are adding. Use this operation with caution, as you might not want to clutter your server space with un-necessary or redundant files.

Check out / Copy

This operation creates a local copy of all the files in the repository to your local machine. Remember, the files thus downloaded are local copies for your use only. Editing them will not affect the master copy maintained in the repository. If you are planning to work on multiple projects simultaneously, you might have to check out multiple copies so that you can work independently.

Edit / Open

Before modifying any file on your workspace, you need to edit / open that file. This operation tells the version control tool that you will be modifying this file locally in your machine. Some tools like perforce won't even allow you to modify a file unless you explicitly perform an edit operation on it.

Update / Sync

Use this operation periodically to keep your local copy synced with the latest version at repository. If you checked out your workspace some time back, and in the meantime some of your co-workers have updated the main repository with new code changes; you will not get the changes in your local copy unless you perform an update or sync operation. Once you sync your workspace, the new changes will be merged seamlessly into your local copy.

Resolve/ Merge

Suppose you have a lot of opened files in your workspace which contain your changes. Next you performed an Update or sync operation. Now, if there are code changes in one or more files at the repository to the files that you have edited locally, they will not be automatically synced. Resolve operation, as the name suggests will properly merge both the changes. At times, you will have to do a manual merge if the tool is not able to do it automatically.

Submit / Check in

Submit or Check-in is the process of sending your code changes to the main repository. With each check-in you might have to specify a comment explaining your changes or the description of the bug which is supposed to be fixed with this check-in. Normally a submit operation creates a changelist with a unique tag / hash key which can be used to identify the check-in in future. Before any submit operation, make sure to update your code to the latest revision and then perform a resolve/sync operation so that you don't overwrite anyone else's changes in the repository.

Revert

Use this operation to throw away your all your local changes in a file and get the latest revision from the repository. Once you revert a file, you need to Edit / Open it again if you wish to modify it. You may also revert to any other version of the file by specifically mentioning it as a parameter.

Diff

Diff operation lets you compare the two different revisions of a file or folder. This is really helpful if you want to understand what all changes went through in a particular changelist. You can also take diff between your local copy and the latest version at repository; this will give you a good view of the changes you have made locally.

Lock

This lets you take control of a particular file and lock it so that no one else can modify it.

Private branch and Merging

Suppose you or a very small group of developers wants to work on some project which is going to take some time and is not aligned with the other work going on for the upcoming release. With this limitation, you cannot do incremental code checkins to the main repository. The main repository is there for the entire team, what you need here is a private repository for your small group or team.

To facilitate this need of a private repository, you can create a private branch and your version control tool will treat it as yet another repository. Remember the Fig 8.1 which represents a simple repository, the same figure can be modified as shown in Fig 8.2 if version control tool is also managing a private branch named – "serverOptBranch".

Different version control tools use different set of commands to create private branches, so please consult the documentation provided with the tool itself. Usually a private branch is created from a particular revision of the main repository. Once a branch has been created, Update and Sync operation will sync the code checked-in into this private branch only and not from the main repository. But, if you feel the need to get the latest version of code from main repository, you need to perform a merge operation. In other words, the code by default shall be synced only from the branch you have created local copy of.

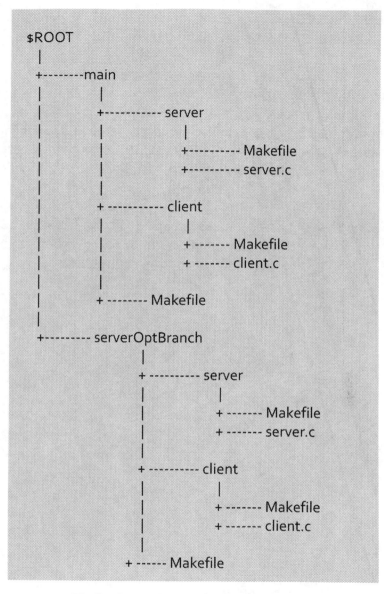

```
$ROOT
   |
   +--------main
   |        |
   |        +----------- server
   |        |              |
   |        |              +---------- Makefile
   |        |              +---------- server.c
   |        |              
   |        + ----------- client
   |        |              |
   |        |              + ------- Makefile
   |        |              + ------- client.c
   |        |              
   |        + -------- Makefile
   |
   +--------- serverOptBranch
                |
                + ---------- server
                |              |
                |              + ------- Makefile
                |              + ------- server.c
                |              
                + ---------- client
                |              |
                |              + ------- Makefile
                |              + ------- client.c
                |
                + ------ Makefile
```

Fig 8.2 Repository with a private branch

Branch Merge operation takes the code from one codebase and integrate the changes to another one. Once the merge operation is complete, you need to resolve the changes too. Merge is done at branch level when you are done with your changes in the private branch and want to submit those changes to the main repository. In this case you

can simply merge the private branch with the main repository and the version control tool will take care of the rest.

Version control tools offer a lot many features but what is mentioned here is mostly that you'll need. You may try to master all the commands and operations provided by the tool being used in the company. It will not only save your precious time but you can emerge as a savior for your teammates in the crunch time. So master the version control tool you will be using in your professional environment because everyone loves a Hero!!

Next, Let's check the second last chapter of the book i.e. Defect Tracking. Buck up, we just have couple more pages to go before you can rightfully claim to be a league ahead of other counterparts.

Chapter Review

- All of us keep a backup of our code base so that if one copy is lost, at least we have the base version. But, when we are not working alone and when everyone has to back up their piece of code, definitely we need a professional tool.
- Version control tool provides a lot many other features along with the basic feature of keeping up a centralized common backup for all.
- Repository is a common place where all the code resides and all of the developers can create a local copy of it in their machine known as workspace.
- Version control tools provide three basic features – parallelism, tagging and reversibility.
- One can easily create a workspace, make changes to code in his local copy independently and then finally submit the changes back to main repository for everyone to see.
- Version control comes from the meta-information kept for every check-in or submits operation.
- Repository can be kept at a centralized location or it can be used as a distributed system. Each has its pros and cons.
- Basic lingo of different version control tools is almost same and includes phrases like – check-in, check-out, sync, update, revert and resolve.
- Private branches can be created if you want to work on some project which should not conflict with the main repository. Once you are done with code changes in private branch, you can merge the branch to the main repository and make the changes available for all.

9

Defect Tracking

If you know about "**Zarro Boogs Found**" you probably know what this chapter is all about. Even if you do, it wouldn't harm to go through the shortest (but quite informative) chapter of the book, right? Here's a brief about the quoted string before we proceed.

Bugzilla, one of the most widely used defect tracking systems returns this error string if the bugs search query returns no results. As Wikipedia mentions, "'*Zarro Boogs' is a facetious meta-statement about the nature of software debugging*". What it infers from the misspelling of 'Zero Bugs' is a friendly reminder that even though the query did not find bugs, your software has them, it's just that they are yet to be discovered.

The job of a Software Engineer is not only about creating the software but also maintaining it till the end of life (either of the software or of his own :P). Software with zero defects is a myth. From testing as well as the usage of software in the field, defects and necessary enhancements are identified. Defect tracking system, as evident from its name is a system used for reporting, tracking and monitoring such reports. These systems are also referred to as bug tracking systems. Few examples of such systems are- *Bugzilla, Teamtrack, Trac, BugTracker.NET, JIRA* and so on.

Components of defect tracking system

A defect tracking system is one of the most important tools used in an organization. Startups with few members may start with maintaining a common excel sheet for the purpose but as the number of employees grow, having a professional defect tracking system in place becomes imperative. The important components of a defect tracking system are as follows:

- The main component is the database of the known issues that contains their details like Bug ID, Title, Description, Identity of stakeholders and so on.
- Other important component is the interface using which the user interacts with the system. These days it is mostly a http browser based interface which is system independent and accessible from anywhere.
- Notification component notifies the stakeholders regarding latest updates of bugs.
- Reporting component generates various reports and metrics that are useful for project management purpose.
- Integration component that integrates the system with external third party systems like Version Control or Project management system.

Screenshot on the next page shows interface of bugzilla displaying Mozilla Firefox crash bugs.

Screenshot of Bugzilla displaying the Mozilla Firefox bugs

A Bug's Life

Being the bright person you are, you already know that this is not about the famous animated movie or insects but what guarantees the job of everyone who is a part of software industry. Jokes apart, different defect tracking systems use different terminologies but more or less they depict the bug's life cycle as follows:

Open/New/Assigned

This is the first stage in a software defect's life. When a bug or enhancement is identified in the software, it is reported into the bug tracker. While submitting the issue, certain relevant mandatory fields need to be specified e.g. Title, Description, Request type, Component, Severity, Priority and so on. Some fields like Assignee or Verifier get auto-filled according to component categorization or can be manually specified. Here is the brief description of the mentioned fields-

- **Request type** - Usually its either Bug request or Enhancement request.
- **Description** - It provides the details of the behavior observed, steps to reproduce the issue or any other specific details required to analyze the issue.
- **Component** - These are pre-defined values specified by the organization. This field helps assigning the issue to correct team or person and also in generating the reports. Components may as well have sub-components defined.
- **Severity** - This is defined in Service Level Agreement as agreed upon by the organization. In simple words, it specifies the impact of the defect on the overall system. Usually there are 1-5 severity rankings, 1 being most severe.
- **Priority** - Some companies maintain this field as well which depicts the priority this issue should be addressed at. However, some companies use the severity field only as more severe means higher priority.

In addition to these primary fields there could be various other fields that are provided by the defect tracking system for monitoring and reporting purposes. You may want to explore more about them on-job itself.

Not an issue/Duplicate

After raising the report, the assignee or someone may review the report and identify if it qualifies to be a new issue or not. The reasons for report being not identified as issue can be something like these:

- If the reported behavior is expected or due to a known defect say a feature is being developed and certain activity is expected to fail till its checked in, that would not qualify as a new issue.
- If the issue is already fixed in the newer builds than the one on which the tests are being performed.
- If there is an existing issue describing the same report or the root and fix of the two reports is same, then the new one can be just marked as duplicate of the earlier reported one.

Before filing an issue, the reporter should make sure that it qualifies to be a new issue. Otherwise, he/she may like to update the corresponding report or update the severity.

Note: While reviewing the report, one may find it related to some other component or needing update in the Severity, Assignee, Verifier field etc. One can update the same any time if they have the credentials to update tracking system.

Fixed

If the submitted report is confirmed to be a new issue, it is worked upon by the assignee and once corresponding changes are checked in, the report is marked as Fixed. The corresponding change details and a brief description of the fix made are updated in the respective fields. It serves as an intimation to the verifier to verify the fix in the newer build.

Verified

The verifier runs tests and checks if the fix made properly addresses the issue. If not, the issue is put back into Open/Assigned state and is assigned back to developer. The verifier shall mention details about tests performed and which all did fail. If the fix passes all the tests run, the report shall be marked as Verified.

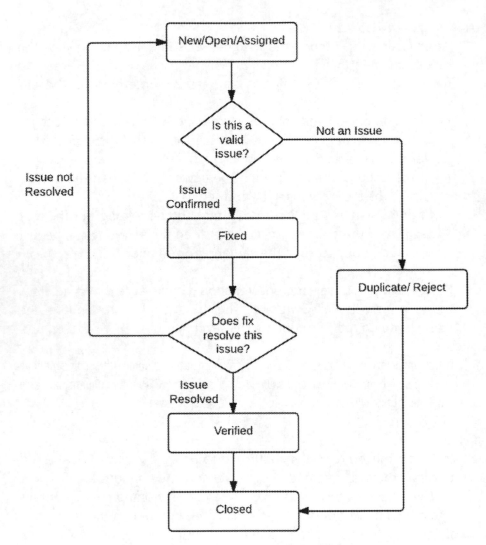

Flow diagram of a Bug's Life

Closed

    After verification, the issue may be moved to closed state. This means the mentioned issue has been successfully resolved and verified. However, If afterwards, something causes the issue to reappear, the report can be brought out of the closed state and marked as Open/Assigned.

Note- Other than the states mentioned above, there is one more state called <u>Monitoring</u>. Sometimes an errant behavior is randomly seen only when an environment reaches some specific unreported stage. An errant behavior would be observed but while trying to reproduce the same in a regular system, it would not occur. In such cases when reason for the issue or verification of the fix could not be confirmed immediately, the report is moved to Monitor state. If the issue is not observed for a certain period, it can be closed or else when its observed next time, its reopened and more carefully inspected. Sometimes, even the assignee is asked to take a close look in the environment itself where it re-appears.

Integration with Project Management & Test Tracking systems

In college, you would have had to pick a project, work on it in the due time and present it to your guide or evaluator. Of course, being the sincere student you are/were, you would have taken the path less traveled and done it all on your own, building it right from the scratch, isn't it? You would also have learnt about various stages of software development in your boring Software Engineering class and while doing your project would have keenly followed the steps like Requirements Gathering, Design, Coding, Testing, and so on, right? ;)

Most probably you might not have needed a process management or test tracking system as the scope and magnitude of the project was not too big or most of the work would happen just on the night before submission and somehow you would present some working piece of code to the evaluator or your guide next morning.

Well, in industry the things happen in bit more structured way because software you write fetches bread and butter instead of grades. Therefore, the projects are tracked and managed so that the committed deliverables can be delivered in stipulated time. What is used for planning, organizing, monitoring and controlling resources to achieve goals within a certain period is called *Project management system*. Ask any Project managers, Team leads or their bosses and they will tell you how their professional life revolves around it.

Similarly, testing is a serious business because an unidentified bug can cause havoc at customer's site and belittle your organization's reputation. For making and tracking various test plans, the tool used is called *Test tracking system*.

Both of these systems also work with bug reports only but more so from Project management and Testing perspective respectively. As these systems pretty much relate to each other, they are now a days either packaged into one single solution e.g. JIRA or provide options to integrate with each other. Worry not if it seems confusing now. It will make more sense on the job.

Hey, by the way, did you realize that you have reached the concluding chapter now. It is about *Product Shipment Cycle*, another topic that would be absolutely new to you and which is why you would definitely enjoy it. Go ahead!

Chapter Review

- A Defect tracking system is required to keep a track of the current outstanding bug right from the stage of filing an issue to the final verification of the bug once a fix is made by the developer.
- Important feature of any defect tracking system is to have all the details of a bug like a unique bug id, who is the developer and tester, steps to reproduce, the release or build on which it is found etc.
- A defect tracking system also provides a reporting and monitoring feature so the progress can be tracked by all and also if the status changes an email would be sent to all of the involved engineers.
- Several stages involved in the life cycle of a bug can be summarized as – open / file a bug, close the bug either by cancelling it or fixing it, verification by test team, reopen or close the bug accordingly if the verification fails or passes.
- To get the best out of a defect tracking system, it can be integrated with other project management tools which make it easy and convenient for managers and others to keep a track of each and every status change and thus make their decisions regarding project status.

Product Shipment cycle

As you make it to the last chapter of this book you would have already understood that excelling in professional software environment is not only about understanding and writing wonderful code but also about understanding and exercising various processes and standards. Last few chapters focused to help you get accustomed with many important tools which you will be using on daily basis.

This chapter discusses the stages or cycles that software goes through from its initial draft till the packaged disk is out in the market. You may not feel the obvious need to go through it in detail, but almost everywhere there is a dedicated team of individuals handling product shipment stages (and they are paid more than the developers!). Directly or not, in your job you will always be involved with this process - your deadlines of deliverables will be governed by the decisions taken by the release and product management team.

Don't worry; it is not as scary as it sounds. Let's walk through the intricacies involved in the release management process in a step by step manner.

Software release
A release in terms of software jargon refers to the periodic distribution of final version of a specific software application. This distributed version

normally includes the relevant binaries, libraries, documentation and packages related to product support. This release is provided to the new customers and also to the old customers if they are willing to continue with the same product. The frequency of distributing the release depends heavily on the customers and the market situation and also it varies from product to product. Some are released as frequently as once in a quarter and some are released once in two years.

On a very superficial level software releases can be classified into two basic categories:

Main release

Also referred to as major release, this is the periodic release stuffed with the latest features, enhancement and other requests. The frequency varies as discussed above based on the situation. This is the release that is going to extend business through new customers and also retain business with old customers. With every major release, customers are usually awarded a support package in the form of a contract. You can understand it as similar to warranty.

For example, a 2 year contract for support and maintenance means that if a customer purchases a license for this product, for the next 2 years he is entitled to claim support for any bugs or issues that would appear at his deployment site(s). The support could be through emails, patch releases with bug fixes or if required, an in-personal visit from the support providers. After two years, he is not liable to the support and he must purchase new license for the latest release version prevailing at that time. A Service Level Agreement (SLA) is agreed upon in the beginning itself which roughly states the agreed upon expected resolution time depending upon the type or impact of an issue.

Patch release

Also referred to as maintenance release, this is generally a follow up of the previous main release with bug fixes and other customer issues resolved. Again the frequency varies based on the product and other factors. But, definitely the frequency of a maintenance release is more than that of main release. At times, when a customer finds a severe issue in the current release which he is using and he wants immediate fix, support teams can provide a patch release almost instantly once the issue is fixed without following any periodic schedule.

Usually patch releases don't account for business as such; they are part of the initial contract/license distributed with the main release. However, the support itself comes with its own price which is included in the major release itself. So in a sense, customer has already paid for the patch works and thus any customer issue gets higher priority than the ongoing work for the upcoming release. So be prepared if you are suddenly asked to divert your efforts to address a customer issue while you were deeply involved in something very interesting and important for next release.

Release management

Release management refers to the process of managing and maintaining the release cycle on track for any software application starting from development phase to the release phase. The ultimate goal of release management is to create a streamlined and effective workflow between developments – build – testing – deployment so as to catch errors as early as possible. Since all of these components are highly dynamic and involves many different people, it makes sense to have a dedicated resource to do the task. Release managers, commonly known as RM caters to this need. They may use release management softwares which are specifically designed to handle all the planning and tracking that is required to get the release out on schedule. Listed following are few of the tasks that are an inherent part of a release manager's job profile:

Track Schedule

Most important job of RMs is to do all the book-keeping to get the release right on schedule. To achieve this they do regular follow-up with managers, developers and testing team through emails, sync up meetings and whatever means they can resort to.

Common environment for dev and devtest

Maintaining a common environment for developers and testers saves a lot of hassles when you have to go a lot of back and forth while working on a critical issue and getting it tested quickly.

Otherwise, this wastes a lot of time in petty issues like tester not able to find a proper environment to test and corollary developer not finding the issue on the setup he is using. Following uniform practices helps understand the progress of tasks even to the ones who are not directly involved such as RMs; they can also easily figure out what's going without bothering the ones who are directly involved.

Nightly builds and automated testing

Working on multi-engineer projects comes with their own side effects. Suppose the release is scheduled to happen by end of this month. All of a sudden there will be more and more checkins going into the main repository and more and more bugs being unraveled by testing team. This asks for more checkins and because of which sometimes more bugs get introduced in a very short span of time just before the release. It will be a nightmare for the RMs and others involved in the release process to find themselves stuck in this cycle.

To catch errors right up front and as soon as they occur, management needs to track every checkin and see if it is causing any issue or not; which again is a cumbersome task. To tackle these issues, software community has come with the concept of nightly builds which means every 24 hour there will be a build generated from the latest code taken from the repository. This ensures that any build breakage is caught quickly and on top of that the automation scripts can figure out the change that caused the failure and send a mail to the respective engineer.

Apart from ensuing that build is fine, scripts can run the whole test suite bundle to check whether any test case failed. This is very useful in case someone forgets to verify all the cases before his checkin. The automated testing can then send the list of failed cases to all the engineers who checked-in their changes in the last 24 hours. Having these processes running every day greatly reduces the risk of unknowingly passing on bugs to the code base just before release. Moreover, based on the results from these scripts developers will know whether they can safely sync their private copies to the latest version from repository or not. Also, RMs can use tracking tools upon the data retrieved from automation tools and thus can decide upon the code release dates because now they know the most stable version of the code.

Archives and report viewing

Running nightly automation is one task and to capture and archive the data reported by them is another. Creating archives of all the stable builds on a central repository is of immense help to all the engineers. Suppose a devtest engineer finds a bug in the latest code, quite naturally he will also want to figure out the build or change which introduced this issue. He might not be having the access to source code, but if there is an archive of all the stable nightly builds somewhere, he can definitely use the previous few versions of the builds to figure out when this bug has slipped in. Similarly, all the reports generated for the test automation can be kept in some repository for later viewing so that everyone knows which build is clean and what kind of outstanding issues are still present in that build.

Develop internal tools

Another job that is part of release management cycle is to develop tools that can greatly enhance the whole process of release management. Such tools involves writing automation scripts for build, scripts for running and adding new test cases, creating a dashboard where all the detailed reports are visible. Apart from this there is a constant need to have the latest release dates updated somewhere for public viewing within the firm and at times outside too. As there could be hundreds of checkins happening every day, efficient tools are also needed to quickly figure out the suspected changes which introduced a recent bug. In short, there is whole set of skillset and dedication required to do the release management task and this is the reason why most of software firms have a separate team to do the job.

Flow of Release management cycle

This section describes the workflow of a software application after the coding stage is over and the release management processes takes over. Long ago during 1950s, IBM introduced a terminology known as alpha/beta for different types of testing stages. Although, IBM dropped the usage of terms themselves; it is quite popular in the current development scenarios. Listed next are few of the important phases and technical terms involved in the process of release management cycle:

Alpha release

It is an internal release within the organization and marks the beginning of first round of testing by testing team. This is done at a time when all the feature development is considered over. It is rarely made available outside the organization as it is considered unstable mainly because it has not been properly tested and thus might give unexpected results at client side.

Code freeze

Alpha release announcement is a two way notification which is meant for both developers and devtest engineers. To the testing team, it means they have to gear up with all their tools and start digging out bugs and other potential issues. To the developer this means that they can no longer work on any enhancement or feature related projects and this particular phase is aptly termed "code freeze". It doesn't mean that developers will be relaxing during this period. In fact this could be worst nightmare of their lives if testing team finds tons of bugs during testing in their code. Thus it is evident that the whole development team will be busy fixing the mess that they have introduced in the code.

Branching out

Once code freeze is enforced and basic testing is over, further feature development is not allowed and the software is marked as "feature complete". A new branch is cut from a stable point and further testing is done on this branch only. This unblocks developers from doing their feature development and at the same time allows them to fix issues raised by testing team in the new branch itself.

Beta phase

Testing done during beta phase is more comprehensive and focus is mainly on the usability of the software at the actual client side. This also includes testing under stress condition and performance analysis of the product as a whole. Beta release is the first release that is published outside of the organization. Beta testers are usually selective new customers eagerly waiting for the release or old and trusted customers who have done beta testing in past too. The bugs found by beta testers needs to be fixed with utmost urgency and care.

Release candidate

With the continuous testing and bug fixes, over a period of time the beta branch will stabilize. This is when it is considered to be worth of a potential final release to everyone. By this stage, everything seems to be in control as all the features are complete, thoroughly tested and passed through one or more beta testing cycles.

Release to Manufacturing

RTM is generally applicable to a software product which is bundled along with another hardware product. This term simply means that the software is ready for its integration with its partner hardware. RTM version of a software is released before it is released to public so that the manufacturers get a chance to test it out within their integrated environment.

General Availability

GA is the final stage in software release cycle where all the activities related to a release process are considered complete. This may include completion of documentation, contacting software distributers, making software media availability either physically or through web and finishing other commercial tasks necessary to release the product in market. Once all goes well, a GA release is made and the software is marked as gone alive.

Project management

There is a separate project management team which works in tandem with RMs to manage the application from a holistic point of view. Like release management team, this team is also not directly involved with development or testing but they are involved indirectly throughout the life cycle of software application. The job of project managers (also known as PMs) is of utmost importance and this chapter wouldn't be complete without a discussion over the several tasks attributed to project management. Following is a short description of the few important phases involved in the overall project management cycle.

Requirement collection

The very first task of any PM is to collect the requirements for the upcoming release. The requirement could be based on the current market, how the competitors are doing and also on what the current customers want next.

Filtering requirements

The list of requirements could be very huge and it is not always possible to meet all of them for the next release. Therefore, filtering is done to decide upon which all would be taken up. The decision to pick or reject certain requirement could be based on several factors such as – importance of customer, revenue that could be generated from the given requirement, available bandwidth of the involved development team and so on.

Distribution of tasks

Once the important requirements are short-listed, they are distributed amongst the relevant development teams. This sometimes involves in-personal meetings with the folks from the development team explaining them the importance of the requirements and the time frame to get it done. The requirements are generally captured at a central location where it can be viewed and updated easily by all.

Track continuously

Once the requirements are in some central repository such as internal web server, it can be tracked easily by PMs. Developers can go on and update the current stage of the project, similarly testing team can also update where it stands in terms of stability. As an engineer, be sure to meet your deadlines because someone is continuously tracking your progress.

Deciding over code freeze

As PMs are constantly tracking the progress of all the projects throughout the organization they are in a better position to decide over the dates for marking the application as feature freeze. Once code freeze is in place, RMs will marshal their troops to do their best to ensure all the processes are followed meticulously.

Beta and main release

PMs are the ones who listed down the important requirement in the initial stage, thus they put a continuous watch on the beta release for the features that are going to be part of it. Same goes for the main release, for they are answerable to higher management for delivering the requirements on time as they are already committed to customers.

Document outstanding issues

Almost always, even the final release is not free of all the petty bugs. All such issues need to be mentioned in the release documents so that customers are aware of them. Apart from this, all the new features and the changes in current ones if any would also make up to the release documents.

Sales hand off

Once all the activities pertaining to the release management are over, PMs will brief up the sales team and make them aware of what this new release means to the customer and the market. This transition is also known as "sales hand off". Now, it is the job of the sales team to entice new customers and explain them the exciting features encapsulated in this new release.

Back to where it all started

As soon as a release is done, PMs start collecting the new requirements and the pending ones which could not make up to this release and the whole cycle repeats itself again.

🔊 Making a good software application isn't only about wonderful coding and rigorous testing; it requires multiple processes to be followed rigidly. Release management and Project Management teams take the burden of doing this task much to the relief of other engineers who can happily remain focused on their assignments.

Chapter Review

- Release management team ensures that the processes involved in software release life cycle are properly followed and for this they may take help of several internally developed tools to facilitate smooth functioning of daily tasks and easy tracking.
- A release can be broadly classified as "main" or "patch". Main release is the major release of an application and it is committed to get new features in and has the ability to generate market revenue.
- On the other hand, patch releases are the support release for the previous main release and a part of license agreement.
- To facilitate smooth functioning and catch errors at the earliest, an automated process builds the latest code and tests it against the set of specified test suites. This can be done every night, every alternate day or even weekly depending on the requirement.
- The flow of release cycle of a software application follows a specific pattern which is – Alpha release, code freeze, branching out, beta release, Release candidate, RTM and finally GA release.
- The job of project management team is to track the project progress right from the initial stage to the final stage of release. The PMs remain continuously in touch with RMs and other teams to achieve the final goal.
- The flow of project management cycle involves several teams but mainly it is the PMs who do all the constant vigil and tracking. The flow pattern is – requirement collection and filtering, distributing the requirements to development teams, deciding over dates for code freeze, keep a watch over features in beta and main release, ensuring that documentation is up to date and finally handing the job to sales team.

Appendix A

Error handling using Try-Catch block in Java

The very basic form of error handling is to free up used resources, display some error message and return back to the caller with proper error code. And obviously it has to be done for each and every error condition in your program. But, writing error handling code along with return statements at multiple places within a function render the code non-readable and hard to debug too.

Chapter 1 discusses in detail about the usage of multiple return statements and how to judiciously use them and still maintain the readability of code. Another elegant solution to this problem is to use "try-catch" block if your programming language supports it. This section gives a basic overview on how to use "try-catch" block in Java. Following is a sample program which tries to access an out of bound array index.

```
1.  public void main(String args[]) {
2.      int[] int_array = new int[10];
3.
4.      try {
5.          for (int i=0; I <= 10; i++) {
6.              int_array[i] = 0;
7.          }
8.
9.          System.out.println("Array initialization
    done..");
10.     } catch(ArrayIndexOutOfBoundsException e)
11.         // exception caught
12.         System.out.println("array accessed out of
    bounds");
13.     }
14. }
```

Java provides an exception handling class which when used with "try" and "catch" constructs enables us to write all the error handling code in a common separate block rather then spread across all over the place.

The compiler in this case throws an error (known as exception) if the code is written within a "try" block. Also, to write all your error handling code Java provides a "catch" block where you can catch specific type of error.

The print statement within the "catch" block will be executed once the index value within the "for" loop reaches 10. Once the exception is caught, execution of the program moves directly to the "catch" block skipping the print statement directly after the "for" loop. If you are a keen observer, you would have started admiring the beauty of "try-catch" blocks. Now you do not need "return" statements scattered throughout the code; as soon an exception is raised, execution of program directly jumps to the proper "catch" block where you can write all the error handling statements. The exception in the above example is thrown by compiler, if you wish to throw your own custom exception check the following example:

```
1.  try {
2.      throw new NullPointerException("the pointer
    accessed is null");
3.  } catch (NullPointerException e) {
4.      // nullpointerexception caught, do your cleanup
    here
5.  } catch (Exception e) {
6.      // other exceptions will be caught here
7.  }
```

The above example also demonstrates that there can be multiple "catch" statements for a single "try" block. However, there is a catch when you intend to use multiple "catch" blocks. "NullPointerException" is a subclass of "Exception" class and hence should be used in "catch" block in the same order. So, if we reverse the exception being caught in the above example, compiler might throw an error as all the exceptions will be caught by "Exception" class and the code within the "catch" block placed for catching null pointer exceptions will never be executed.

Along with "catch" block, Java also provides another construct – "finally". If there is no exception caught by "catch" block, "finally" gets executed. Also, if an exception is caught then the "finally" block gets executed after the relevant "catch" block. Only catch is – code in "finally" block should be exception free. So, whatever be the case you can always assume to write error handling statements in "finally" block. Following is an example of its usage:

```
1.  try {
2.      throw new NullPointerException("the pointer
    accessed is null");
3.  } catch (NullPointerException e) {
4.      // nullpointerexception caught, do your cleanup
    here
5.  } finally {
6.      // this block is executed regardless of exceptions
    being thrown
7.  }
```

Appendix B
Installing open source software distribution

Throughout this book, you would have come across many open source softwares. The beauty of open sources bundles lies in their free of cost availability, but you need to follow few simple instructions to install them and get them working. This section touches upon the installation guide so that you can readily install almost any of the freeware on Unix related operating systems. Although the example taken here is that of all powerful editor – vim, but you should be able to apply the same steps to any open source bundle.

Best way to install vim is to download its source code from http://www.vim.org/sources.php or any other genuine resource. The source code bundle is available in zipped format such as "tar.bz2", "tar.gz" etc. You can also get source code through version control systems like CVS or SVN which lets you remain updated to all the latest changes made by developers throughout the globe.

Unzip the downloaded bundle
Once you have downloaded the source code, the first step to step forward is to un-compress the bundle. For this you would require "tar" utility which usually comes as an operating system utility.

Configure to build
Once the source code is uncompressed, change directory to the location where it is extracted. Usually "tar" extracts the zipped files under the current directory or a new directory under the current directory. Also, the extracted files maintains the proper directory

- Uncompressing tar.bz2 file
  root@freebsd # tar xvfj vim-7.3.tar.bz2
- Uncompressing tar.gz file
  root@freebsd# tar xvfz vim-7.3.tar.gz

x – extract
v – verbose mode output (display files as they are extracted)
j – file is bzipped
z – file is gzipped
f – read from a file

structure which was present at the time of zipped file creation. If you glance through the extracted files, you will find README file which contains detailed instruction to proceed forward. There will be a file named "configure" containing several parameters and their values which are needed to configure the build process. Few of the parameters include the name of the compiler to choose, path to other dependent libraries, flags to specify whether a particular feature needs to skipped, path to hardware related dependencies etc.

You should be fine even without an understanding of the "configure" file, but if you get some errors during compilation then you might have to look at the configure file and figure out what exactly is missing.

```
1.  $ ./configure
```

Compile Source code

Once you have run the "configure" executable, it will create a Makefile specific to your local platform. Simply compile the code using "make" utility.

```
1.  $ make
```

Install the Sofware

You are almost there if compilation succeeds without any error. Once compiled, the binaries and libraries will be copied to "bin" and ".libs" directories respectively in the current directory. You can use them as it is, however you will always have to access the vim executable with full path e.g. */var/mydir/vim-7.3/bin/vim.*

In general, all the binaries are either present in */usr/bin* or */usr/ local/sbin* and this path is included in $PATH as shell variable so that to access any binary present in above location you need not give full path qualifier. So, to get your newly installed vim binary to be accessed without full path you can either add */var/mydir/vim-7.3/bin* to $PATH or copy the binary manually to */usr/bin.* There is a neater way to achieve the same using "make" itself. Use the following command and it will copy the binary to */usr/local.*

```
1.  $ make install
```

If you want your binaries to be installed in some other location, you need to pass on the directory path while configuring the build process using "configure" utility as discussed above. Just modify the **PREFIX** parameter in the "configure" file itself or provide it as a command line argument to the "configure" utility. By default **PREFIX** points to */usr/local* in the configure file.

```
1.  $ ./configure -prefix=/var/myvim
2.  $ make
3.  $ make install
```

Following above instruction will compile and finally copy the relevant binaries to */var/myvim.*

So, that is how easy it is to install any open source software. Have Fun !

Appendix C
Integrating cscope with vim

Cscope is a wonderful tool providing developers the ability to search C symbols throughout the complete source code efficiently. The interactive interface of Cscope is good enough to love the tool, but at times it becomes cumbersome if you are busy working in vim and all of a sudden you have to switch your view to use Cscope. However, this problem shouldn't trouble you by any means because Cscope can easily be integrated with vim so that you can use all the functionality of Cscope while still residing in your code browser view. The following discussion lists down the steps that you should know in order to complete the integration. Also, note these integration steps applies to vim-6.x or later.

Get a copy cscope_maps.vim file

Before you start, you must download your copy of cscope_maps.vim from:
http://cscope.sourceforge.net/cscope_maps.vim

Then copy this to your home folder under $HOME/.vim/plugin location. Also, your vim should have been compiled with "--enable-cscope" earlier, otherwise you will have to configure and compile again.

Setup initial shell scripts

Once you have copied the map file to plugin directory, make it available at the shell startup so that you don't have to set it up every time you open a new shell. For bash, add the following lines in your $HOME/.bashrc or $HOME/.bash_profile file:

```
1.  PATH=$PATH:$HOME/.vim/plugin
2.  EDITOR=vim
```

That's it, and now if you open a vim session from the directory containing your Cscope database, you can use Cscope from vim. Move your cursor over to a C symbol and press "CTRL+\ s" and you will see all the places this symbol is used. Choose one of them by pressing enter and the same will be opened in present vim window only. Type "CTRL+t" to go back to old file. Type ":help cscope" to learn more about using Cscope from vim or go through all the shortcuts present in cscope_maps.vim file itself or learn it through any genuine resource on internet.

Setting up CSCOPE_DB

Setting up map file and .bashrc lets you use Cscope within vim but still there is a limitation to this approach. You always have to open vim session from the directory where your Cscope database file – "cscope.out" lies. To circumvent this problem, we need to setup the database path in some .rc file. Also, to get this to work properly the cscope database should be built with absolute path for the involved files. For bash, add the following set of commands to your $HOME/.bashrc or $HOME/.bash_profile:

```
1.  cd /
2.  find /var/myproj -name "*.c" -o -name "*.h" >
    /var/myproj/cscope.files
3.  cd /var/myproj
4.  cscope -b
5.  CSCOPE_DB=/var/myproj/cscope.out
6.  export CSCOPE_DB
```

The last export statement is what makes it working. Every time a new bash session is invoked, path for the Cscope database is being setup using variable CSCOPE_DB. Thus, whenever you start a new vim session the database parameter is set and vim knows from where to fetch the Cscope information.

Create an alias to build Cscope database

Another useful tip on using Cscope would be to put the above statements within a function and create an alias so that you don't have to manually type all those commands which can be simply done by executing a simple command alias.

```
1.  function build_cscope () {
2.      cd /
3.      find /var/myproj -name "*.c" -o -name "*.h" >
    /var/myproj/cscope.files
4.      cd /var/myproj
5.      cscope -b
6.  }
7.  alias buildcs='build_cscope'
```

Index

Suggested Reading

Learning should never stop and the faster you want to grow as a professional, the more you should continue to learn. Most of the people think and will tell you that on-the-job learning is enough. Don't fall for that! An average software developer reads less than one professional book per year. You should strive to be better.

Here is a recommended list of some books for further study that have been studied and found quite useful by the authors of this book. This list is not specific to a certain technology or language. However, the learning from these shall be useful for you as your work deals with code. We haven't mention those books which we believe you would have already gone through in college, for example, *The C Programming Language* by Kernighan and Ritchie and *Introduction to Algorithms* by Cormen, Leiserson, Rivest & Stein (Go, get them if you haven't!).

- Programming Pearls Second Edition by Jon Bentley
- The Pragmatic Programmer: From Journeyman to Master by Andrew Hunt and David Thomas
- Code Complete 2 by Steve McConnell
- Head First Design Patterns by Elisabeth Freeman, Eric Freeman, Bart Bates and Kathy Sierra
- Design Patterns by Erich Gamma, Richard Helm, Ralph Johnson and John Vlissides
- The Mythical Man Month by Frederick P Brooks Jr
- Refactoring: Improving the Design of Existing Code by Martin Fowler, Kent Beck, John Brant, William Opdyke and Don Roberts
- Rapid Development by Steve McConell
- Agile Software Development by Robert C. Martin
- Structure and Interpretation of Computer Programs, Second Edition by Harold Abelson, Gerald Jay Sussman and Julie Sussman
- Code: The Hidden Language of Computer Hardware and Software by Charles Petzold